HARVARD ECONOMIC STUDIES

PUBLISHED UNDER THE DIRECTION OF
THE DEPARTMENT OF ECONOMICS

VOL. XXIII

16 Randolph November 23d 1807
    Jonathan Hill Dr to Eben Belcher

by 1 pare of women lat[her] s/3	0 0 71
25 by Cash 3/	0 0 50
30 by 1 pare of stats 1/	0 0 17
...	... 46
... pare of men ... other 4/	0 0 67
...	0 1 ...
... 14/1/500	1 0 0
1808 ... shoes	0 0 16
March ...	0 0 20
	1 2 1 5
April 1, 1 pare of men taps 1/9	0 0 29
9 to 1 pare of men lather calfskin	0 0 92
...	0 0 18
May 2 to ditto	0 0 25
25 to Cash 4/6	0 0 79
June 3 to Cash 6/	0 2 92
10 1 pare of oven lather 5/6	1 7 50
June 2d to Board for hid to June the 20d	19 73
to Cash	0 5 94
	25 89

June 20, 1808, this day setled all accounts with Eben Belcher and made an even balens by reeceeving one Dollar in full of all accounts.

for the
Eben Belcher

A PAGE FROM A DOMESTIC SHOEMAKER'S ACCOUNT BOOK

THE ORGANIZATION OF THE
BOOT AND SHOE INDUSTRY
IN MASSACHUSETTS BEFORE 1875

BY

BLANCHE EVANS HAZARD
PROFESSOR OF HOME ECONOMICS
IN CORNELL UNIVERSITY

TO THE MEMORY
OF MY
FATHER AND MOTHER

PREFACE

INTRODUCTORY SURVEY OF MATERIALS, METHODS AND SCOPE OF THIS STUDY OF FACTS CONCERNING THE ORGANIZATION OF THE BOOT AND SHOE INDUSTRY IN MASSACHUSETTS BEFORE 1875

THE development of the boot and shoe industry of Massachusetts proves to be an interesting and productive field for economic investigation, not merely because its history goes back to colonial days as one of the leading industries of the states, but more especially because the evolution of industrial organization finds here an unusually complete illustration. The change from older stages to the modern Factory Stage has been comparatively recent, and survivals of earlier forms have existed within the memory of the old men of today. Sources, direct or indirect, oral and recorded, can be woven together to establish, to limit, and to illustrate each one of these stages and the transitions of their various phases.

MATERIALS, AND DIFFICULTIES IN COLLECTING THEM

The materials used as the basis of the conclusions given here have been gathered at first hand within the last ten years,[1] by the writer, in the best known shoe centres of Massachusetts, *i.e.*, Brockton, the Brookfields, the Weymouths, the Braintrees, the Randolphs, and Lynn. The collection and use of such written and oral testimony has been attended with difficulty. No New England shoemaker of a former generation has dreamed that posterity would seek for a record of his daily work.[2] Only inad-

---

[1] From 1907-1917.
[2] Exceptions to this did not occur until about 1880, when David Johnson of Lynn, and Lucy Larcom of Beverly, began to write in prose and poetry about the shoemaker's homely daily life. Since then newspaper reporters have become active in getting such dramatic details as the Blake-McKay and the Howe-Singer relations afforded, or the rise of Henry Wilson, of Natick, from the cobbler's bench to the vice-presidency of the United States.

vertently have the pages of account books, kept to help in "settling" with neighbors and customers, bills, and letters giving amounts and addresses, been left for us. These records, however, are scarce and fast becoming scarcer. The spring and fall house cleanings and the demolition of old shops to give place to new factories are destructive to valuable papers of this kind, and even where an old attic still harbors them, oftentimes molding and fading under leaking roofs, they are generally difficult of access. The recollections of old men and women who have worked on shoes, in producing or distributing them, are even more precarious. Such people are often confused by specific questions, even when the investigator is not a stranger to them. Their confidence can, as a rule, be restored only when the appearance of any special course of questioning has been abandoned. The material which is thus gathered comes imbedded in unrelated matter with little perspective, and with much provincialism; yet it is the most valuable, as it is the most vital that can be secured. I have prosecuted such inquiries among the older inhabitants in a number of Massachusetts towns, not only where the industry now exists, but where scattered shops, the " ten-footers " of the domestic period of shoemaking, are the only visible remains of a once flourishing local trade.

Besides this oral and written evidence, there are available also public official records, like town reports, parish books, and custom-house papers; newspaper files with the advertisements of auction sales of boots and shoes, of domestic and imported hides, of demands for apprentices and journeymen, of sailing dates for vessels, and of western despatch or express companies; centennial and anniversary addresses,[1] and occasional local newspaper or trade paper reports of old men's recollections.

[1] For example, a dedicatory address delivered by Mr. Loren W. Puffer at the Plymouth County Court-house, by giving facts about Shepherd Fiske, the agent for Governor Bowdoin of Massachusetts, whose blast furnace at Bridgewater, in Plymouth County, supplied cohorns and grenobles for use in the colonial wars as well as cast-iron kitchen implements for local consumption, furnished substantial clues for a study of the iron business in that locality now so famous for its shoe industry. Upon investigation, an explanation was found for the tardy development of the Domestic Stage in the local shoe industry. The small shops, where nails and other

The contemporary official documents and newspapers give little of direct value, but serve to give the general setting of the community's industrial activity. The local addresses and reminiscences are nearly if not actually contemporary. Though tinctured and limited by personal and local pride, by inaccuracy, and by fallacious reasoning as to cause and effect, they are full of suggestion, and help in the appreciation of facts found in other places or other relations.

SCOPE

The information thus gathered seems, on analysis, to confirm inductively and with definite evidence of the transitions, the stages of evolution set forth by Karl Bücher.[1] In the boot and shoe industry of Massachusetts before 1875, four stages of production may be definitely traced. Although the stages are distinct as to characteristics and essential features, they are not so as to time, for overlaps and survivals occur. The household economy or *Home Stage*, for instance, characteristic of frontier conditions, was early followed by the *Handicraft Stage*, which prevailed until the middle of the eighteenth century. The *Domestic Stage* of industrial organization, with its successive and overlapping phases, was well under way before the Revolution, and lasted to the middle of the nineteenth century, giving place, about 1855, to the *Factory Stage*, which passed into its second phase by 1875.

ACKNOWLEDGMENTS

Neither the research work just described nor this constructive use of the materials thus gathered could have been accomplished without the constant and unstinted aid and the inspiration of Mr. Edwin F. Gay, Dean of the Harvard Graduate School of Business Administration. To him and to the score of octogenarian

---

small iron pieces were made by domestic workers and their apprentices, gave way to the shoemakers' " ten-footers " only when the iron industry had developed stronger competitors and advanced to the Factory Stage. Even then, the tool making " habit " survived and North Bridgewater (Brockton) was famous for its mechanical inventions and manufactures for the shoe industry before it was known as a shoe making centre.

[1] Bücher, Karl: Die Entstehung der Volkswirtschaft.

shoemakers who have given generous help, the author gladly acknowledges her indebtedness and gratitude. To Professor Albert B. Hart of Harvard, who gave to the author her fundamental training in the use of historical sources which made this research possible, like acknowledgment is due.

This manuscript in going through the press has had the painstaking care and interest of Professor Charles Hull of Cornell University.

CONTENTS

CHAPTER		PAGE
I.	HOME AND HANDICRAFT STAGES	3
II.	DOMESTIC STAGE. PUTTING-OUT SYSTEM, 1760–1855. PHASE 1, 1760–1810	24
III.	DOMESTIC STAGE. PHASE 2, 1810–1837	42
IV.	DOMESTIC STAGE. PHASE 3, 1837–1855	65
V.	FACTORY STAGE. PHASE 1, 1855–1875	97
VI.	THE HUMAN ELEMENT IN THE BOOT AND SHOE INDUSTRY	127

APPENDICES

APPENDIX		
I.	PROCESSES ON SHOES IN A MODERN FACTORY	159
II.	MODERN SHOE REPAIRING	168
III.	MEDIAEVAL SHOEMAKING TOOLS	169
IV.	PARTIAL CONTENTS OF THE DELIGHTFUL, PRINCELY AND ENTERTAINING HISTORY OF THE GENTLE CRAFT	170
V.	A CONTEMPORARY ACCOUNT OF NEW ENGLAND TRADES IN 1650	170
VI.	THE CHARTER OF THE COMPANY OF SHOEMAKERS, BOSTON, 1648	171
VII.	EXCERPTS FROM THE TIMOTHY WHITE PAPERS	173
VIII.	EXCERPTS FROM THE GEORGE REED PAPERS	173
IX.	SOME TYPICAL DETAILS OF THE EARLY POLITICAL, RELIGIOUS, AND FINANCIAL HISTORY OF FOUR MASSACHUSETTS SHOE CENTRES	176
X.	EXCERPTS FROM THE SOUTHWORTH PAPERS	183
XI.	EXCERPTS FROM THE BREED PAPERS	185
XII.	NEWSPAPER ADVERTISEMENTS FOR THE SHOE INDUSTRY	188
XIII.	NEWSPAPER ADVERTISEMENTS — TANNERIES	190
XIV.	EXEMPTION FOR SHOEMAKERS DURING THE REVOLUTION	192
XV.	EXCERPTS FROM THE WENDELL PAPERS	193

CONTENTS

APPENDIX	PAGE
XVI. EXCERPTS FROM THE BATCHELLER PAPERS	200
XVII. PAGES FROM THE BATCHELLER ACCOUNTS	202
XVIII. THE BARBER STATISTICS OF MASSACHUSETTS TOWNS	207
XIX. PAGES FROM THE HENRY WILSON ACCOUNTS	213
XX. EXCERPTS FROM THE ROBINSON & COMPANY PAPERS	219
XXI. NATICK STATISTICS	220
XXII. EXCERPTS FROM THE HOWARD AND FRENCH ACCOUNTS	221
XXIII. EXCERPTS FROM THE GILMORE ACCOUNTS SHOWING BANKING FACILITIES	230
XXIV. EXCERPTS FROM THE KIMBALL AND ROBINSON PAPERS	231
XXV. EXCERPTS FROM THE TWITCHELL PAPERS	236
XXVI. THE MCKAY MACHINE	245
XXVII. EXCERPTS FROM THE BATCHELLER ACCOUNTS	246
XXVIII. SAMUEL DREW, THE SHOEMAKER METAPHYSICIAN	256
XXIX. CAREY, AN ENGLISH BAGMAN	258
XXX. AN ENGLISH STORY OF THE BIRTH AND TRAINING OF ST. CRISPIN AND HIS BROTHER	258
XXXI. A FRENCH STORY OF THE BIRTH AND TRAINING OF ST. CRISPIN AND HIS BROTHER	261
XXXII. THE PIOUS CONFRATERNITY OF BROTHER SHOEMAKERS	262
XXXIII. THE SHOEMAKER'S GLORY, A MERRY SONG	264
XXXIV. BROTHERS OF ST. CRISPIN	265
XXXV. ATTITUDE OF CRISPINS TOWARD NEW HELP	266
XXXVI. RITUAL OF THE CRISPINS	266
XXXVII. THE CRISPINS AT BURRELL AND MAGUIRE'S FACTORY	267
SOURCES	268
INDEX	273

ORGANIZATION OF THE BOOT AND SHOE
INDUSTRY IN MASSACHUSETTS
BEFORE 1875

CHAPTER I

HOME AND HANDICRAFT STAGES

Introduction:
    Essential tools.
    Essential processes.
Home Stage:
    Phase 1. Purely " home-made " boots.
    Phase 2. Itinerant cobblers' work.
Summary of the Home Stage of the Boot and Shoe Industry.
Handicraft Stage:
    Phase 1. Bespoke work.
    Phase 2. Extra sale work.
Brief Survey of the General Economic Conditions of early Massachusetts in which the Home and Handicraft Stages of the Shoe Industry were Developed.
Randolph, Brockton, Brookfield, and Lynn.

Essential tools

IN a modern shoe factory, there are one hundred separate operations performed in making one shoe.[1] So many parts and such intricate machinery are deemed proper, if not necessary, for even a cheaply made shoe today, that it is difficult for a layman to understand the technique involved, much less to undertake, as his great grandfather did, calmly and yearly, to make a shoe. Shoemaking, however, can be and used to be a very simple industry.

The tools and processes of shoemaking in all countries and ages prior to 1830 were few and easily mastered. Skill and good materials made excellent shoes fit for a princess; but the same tools and processes were useful for making the crudest shoes such as mediaeval serfs wore. From time immemorial, there have been

    2 parts to a shoe: an upper and a sole;
    4 processes in making a shoe: cutting, fitting, lasting, bottoming;
    8 tools necessary for making a shoe: knife, awl, needle, pincers, last, hammer, lapstone, and stirrup.

[1] Just as a matter of contrast and not with the need of real comprehension on the part of the reader, those one hundred operations are given in Appendix I, and a description of a modern shoe repair outfit in Appendix II, in order to show

Essential processes

These four processes could be performed adequately with just those eight tools by any frontier farmer in his colonial kitchen. From 1620 up to 1830, there were no machines for preparing leather, nor for making shoes in Massachusetts, or anywhere. A lapstone and a hand-made hammer were used for pounding the leather; a single knife [1] for cutting both sole and upper leather. An awl to bore holes, and a needle or a bunch of bristles were necessary for sewing the shoes. This process was called fitting, and consisted of sewing the parts of the upper together. When the upper was fitted, it was slipped on the last, which had an insole tacked to it, and its lower edge pulled over this wooden form tightly with pincers until it could be fastened temporarily with nails. Then the outer sole was either sewed or pegged on to this lasted upper. The last in the shoe was meanwhile held firmly in place by a strap or stirrup, which passed over it and down between the shoemaker's knees where the shoe rested, and was held taut under his left foot.

Home Stage

For the details of the simple processes of making boots and brogans in the Home Stage of the Boot and Shoe Industry of Massachusetts in the seventeenth and eighteenth centuries, we have a rare but satisfactory store of sources made possible by the combination of survivals and the good memories and keen interest of men living in the first decade of the twentieth century. Their recollections have been based upon childhood experiences before the middle of the nineteenth century, in isolated communities in

the advance in technique, the increase of complexity, and the division of labor which have all been made in order to save human labor and to secure uniformity in shoemaking.

[1] The well-to-do particular shoemaker of the same period might, however, have a whole series of knives sticking along the edge of his bench, right at hand to suit his several needs in cutting thick or thin, soft or tough leather, a flat surface or the rounding edge of a sole. One of them he used as a skiver to thin off the leather, one for a heeler and another for paring. See Appendix III for mediaeval tools mentioned in St. Hugh's Bones.

New England, where the industrial conditions and customs, typical of previous centuries, had survived.

Phase 1. Purely home-made boots

From their accounts,[1] it is possible to construct a mental picture of the way the country boy stood on the bare kitchen floor or on a paper and had the shape of his foot traced with charcoal or chalk. Sometimes he merely had its length marked, and he watched his father, looking over the family's meagre collection of lasts with breathless anxiety, lest he could not find a last that came even somewhere near that measure. By the time fall had come bringing frost and frozen ground where the boy was doing his farm chores, this barefoot son was vitally interested in having *some* shoes made without delay.

There was little to be decided after the last was found. The shoes would be either high boots or brogans, as low shoes were then called. They would be made of roughly tanned cowhide and be black or russet in color.

Phase 2. Itinerant cobblers' work

Old men have remembered also a second phase of the Home Stage of shoe production, where the itinerant cobbler appeared, going from house to house with his kit of tools and a few lasts rolled up in a leather apron which was slung over his back, or trundled in a wheelbarrow along with his cobbler's bench. From stories told me by other old people,[2] I have realized how much

---

[1] Mr. Jerome C. Fletcher, who was born and brought up in Littleton, New Hampshire, and Mr. Hiram S. Worthley, of Strong, Maine, have been especially helpful in their reminiscences. They have unconsciously illustrated, by moving their hands and feet to suit the action while they talked, and by imitating the tones which memory recalled, the very methods and atmosphere of shoemaking in their New England homes. They have made the listener feel their childhood's expectancy and keen interest in the annual pair of new boots. Mr. Worthley used to seem to see spread before him, as he talked, the family supply of leather. "There is the hide of one cow for top leather, the hide of one ox for sole leather, and one calf skin for best shoes, and sheep skin for their trimmings."

[2] Miss Abby Cole, of Warren, R. I., and Mr. Benjamin Martin, of Bristol, R. I., have told me that in their community as late as 1821, shoes were made in their houses each year by Gardner Sisson, of Touissit. They owned that he was rather

more desirable it was considered to have the family's shoes made by such an itinerant cobbler, rather than by the father and older brothers. This cobbler was either a journeyman, " whipping the cat " after his apprenticeship to some master in a larger town was completed, or a self-taught farmer of their own community, who could make more at this trade than at farming. His standard was apt to be higher, his experience wider, his number of lasts greater, and his knowledge of leather deeper than that of any other farmer in the village. The second phase of this Home Stage had its difficulties, however, for the delays filled by impatient waiting for the itinerant cobbler sent some farmers in despair back to trying their own unskilled hands at shoemaking to meet the family needs, though they knew that the shoes they now turned out must suffer comparison with those made in the neighborhood by real shoemakers.[1] Incidentally they thus kept alive the first phase of shoemaking.

The head of a family had not only to use foresight in engaging betimes the cobbler, as well as the tailor, but he had to forecast also his own demand for leather. Every community had its bark house and tanning pits, where either hemlock or oak bark was thrown in to cover and to cure the hides. The process of tanning seems to have been very simple then, for into the rough pit, dug in the corner of the farm, was thrown a layer of skins, then a layer of hemlock bark, another of skins, and more hemlock bark. For sole-leather they used oak bark, if they could get it, rather

out of style in 1821 and that people knew they could go to Providence, less than twenty miles away, and buy store shoes, but they liked to have Sisson come to the house because he made shoes so well. They told me the delight that they, as children, experienced in seeing grandfather take out the shoemaker's bench from the attic, place it near a window selected with reference to good light in the kitchen, and then in seeing Sisson unwrap the tools which he had brought in his leather apron.

[1] " The Answer to Abiel Kingsbury's Prayer," a story by Miss Virginia Baker in the *Atlantic Monthly* for June, 1913, is an intensely interesting realistic account of a family, waiting for the coming of a cobbler to the vicinity of Rehoboth in southern Massachusetts. Several children were out at toes and Abiel, a widower, distracted by the attention of a maiden lady who wanted to take care of him and mother his children, had not had enough foresight to hire the cobbler ahead. At last he turned to the old Indian woman in the Hollow and engaged her to make moccasins for the children, but decided to hold out for himself until the itinerant cobbler came.

than hemlock bark. Once a year the vat was opened, and the bark and leather taken out. The owner of the vat was currier also, so that he took the hair off the skin, and dressed the leather by pounding it on a lapstone and scraping it. He kneaded with oil the skins used for upper leather, and, in general, he prepared it by the same crude methods which primitive people of all times and centuries have employed.

At first the man who owned the tanning pit used to tan skins just for the immediate use of his own and his neighbor's family. One can easily see the chance for growth of this industry until the tannery, begun by the farmer-butcher-currier as a by-industry to save waste products, or as a mere convenience for the neighborhood, became more important than the business of carrying on the farm itself. As real craftsmen set up in business in the town to do shoemaking under the second or Handicraft Stage of the shoe industry, they probably demanded a more regular supply and a higher standard of leather than the old, occasional by-industry had secured for the itinerant cobbler. The shoemakers who had come from larger towns, especially seaports, were used to working on a certain amount of imported leather and would be critical customers of the village tannery product. It is rather hard to decide whether a successful tannery was accountable for the superior development of a local shoe industry in various New England villages, or vice versa. The fact was mentioned sometimes in town records when a newly arrived shoemaker had come from a community which was known for its tannery.[1] Tanning and shoemaking were both clearly rooted as by-industries at least in every New England village before the earliest colonial conditions of shoemaking, *i. e.*, the Home Stage, gave way to the Handicraft or Masterworkman Stage.

SUMMARY OF THE HOME STAGE OF THE BOOT AND SHOE INDUSTRY

During the Home Stage in the shoe industry in Massachusetts, shoes were made only for home consumption. There was no

---

[1] Cf. p. 22 for the coming of Ezra Batcheller to North Brookfield from Grafton, Massachusetts, in 1810. Note also the mention in the Suffolk Registry of Deeds of the trade of the man who bought land, if he were a cordwainer or a currier, a yeoman or a cooper, a gentleman or a merchant.

market for them. The standard was "the best you could make or have." The farmer and his older sons made up in the winter around the kitchen hearth the year's supply of boots and shoes for the family out of leather raised and tanned on his own or his neighbor's farm. This did not, or could not happen, until the shoes worn over from England were past repair, and until in his frontier life the colonist had secured a permanent cabin and had raised some cattle whose hides might be tanned. How much the individual man, woman, and child suffered from the change to the rough home-made shoes we shall probably never know, and relatively the item of the discomfort of a pair of shoes must have been slight. One can easily imagine the wife or daughter, and even the grown son being told to stop grumbling over a poor fit, and to be thankful that he had any shoes at all.

Handicraft Stage

Phase 1. Bespoke work

The transition from the Home to the Handicraft Stage came gradually not only in each town, but even in the experience of a single shoemaker. One day he might work as an itinerant cobbler on a wage in a farm house, and the next, he might work in his own house, for a bargained price, on shoes for a customer, who might or might not supply the leather, but in either case, agreed in advance upon a price for the product turned out by the shoemaker. Thus the Home Stage, with its chief characteristic of production *merely for home consumption*, gave way to the Handicraft Stage with its characteristic of *work done for a market*, on the specific demand of a *definite customer*. Such work came to be called "bespoke work." Though the future owner of the shoes could no longer have oversight of the worker's use of the leather he had brought to the shop, the standard of work was probably higher in most cases.[1]

---

[1] Mr. John R. Commons called attention in his "American Shoemakers" (p. 42, Q. J. E., November, 1909), to the fact that the master shoemakers by working in shops instead of private kitchens were open to the inspection of fellow craftsmen, as the itinerant cobbler had never been, hence higher standards could be maintained.

HOME AND HANDICRAFT STAGES

The official records [1] of Massachusetts reveal some of the struggles and grievances of early shoemakers and master craftsmen at this time of transition. They wished to cease being itinerant cobblers at the farmer's beck and call,[2] and to work in their own shops on leather which they provided, and at a price which they themselves set. A compromise was made by the legislators whereby the master could stay in his own shop, but he had to make up the leather that was brought by the customer who wished to provide his own, and he had to work at a "fair price." This supposedly was to be set by public opinion which shoemakers and customers alike had a chance to make.

In thickly settled communities this Handicraft Stage in its first phase came sooner and also passed out sooner, especially in the seaports around Massachusetts Bay. While we have seen it surviving in the isolated back water country [3] even past the middle of the nineteenth century, it had passed out in Lynn, the Weymouths, and other eastern Massachusetts towns, nearly a century earlier. In those same places, it had come also a century earlier, for as early as 1654 when Edward Johnson published his "Wonder Working Providence of Sion's Savior," [4] he recounted shoemaking among the numerous crafts being plied in eastern New England. In his account we see an array of craftsmen doing their special work as artisans and depending upon farmers for the food, which would come directly or indirectly, as pay for their services, as early as the middle of the seventeenth century in a few of the older and more urban communities. It was not till the middle of the eighteenth century, however, that the Handicraft Stage was prevalent in the shoemaking industry in New England.

A mediaeval shoemaker would not have found a Massachusetts shoemaker's shop of that date very strange in itself. The master's

[1] See Appendix V and VI for a detailed quotation from these records.

[2] See the story of Paul Hathaway in chapter VI, p. 134.

[3] Cf. mention of Littleton, New Hampshire; Strong, Maine; and Frenchtown, Pennsylvania in this text, pp. 5 and 178.

[4] Cf. Appendix V, for quotations from Edward Johnson's "Wonder Working Providence of Sion's Savior in New England." Edition published in 1654, p. 207 of Book 3.

kit, materials, and helpers would have all been familiar in both kind and number. But the absence of gild officers and regulations would have seemed most unnatural, for there were no gild regulations in these shoe shops of colonial Massachusetts except in Boston.[1] The gild organization was rare enough in the English colonies to be exceptional. Philadelphia and Boston, receiving more new European shoemakers, and being trade centres from the earliest times, developed and maintained it. Public opinion, added to experience gained by watching the actual wearing qualities of shoes, made standards for shoemaking grow higher as community conditions became less isolated and farmers' purses showed the effects of industrial prosperity. Then, too, there was the spur of competition on workmanship imposed upon the workmen by the European-made shoes which never ceased to be imported by the well-to-do, and to be brought over by the newcomers. To do the custom work for the squire of the town meant in colonial Massachusetts what it did to make shoes for the princess in mediaeval France. An apprentice lad in Boston probably dreamed of making and trying on shoes for the Governor's daughter in much the same state of dazed expectancy as the shoemaker of fiction, known as the disguised prince apprentice of Kent, experienced when he made shoes for Ursula, the Roman Emperor's daughter.[2]

Working with the custom shoemaker in his shop, there were generally two or three apprentices and two or more journeymen. The former were bound for a term of seven years just as lads had been in Europe since the fourteenth century. They had to help the dame of the household as well as the master of the shop, and doubtless sang the same refrain apprentices used to sing in England.

> However things do frame,
> Please well they Master, but chiefly they Dame.

[1] Cf. the Charter of Boston Shoemakers, published in full in the records of Massachusetts Bay in New England, vol. iii, p. 132 and printed as Appendix VI of this treatise.

[2] For the details of this shoemaker's tale, see The Delightful Princely and Entertaining History of the Gentlecraft of Shoemakers, pp. 57–73. Published in London, 1725. See also Appendix IV.

They longed for the time when they, like the journeymen they had seen come and go at their master's shop, would " shroud St. Hugh's Bones [1] in a gentle lamb's skin " and start on their travels earning their first money by " whipping the cat " as itinerant cobblers in lonely country houses until they found regular work in some master's shop. All through both phases of this Handicraft Period, the custom shoemakers taught their apprentices faithfully to master all the processes and to make the whole boot or shoe. No suggestion or even prophecy of division of labor, such as became characteristic in the third or Domestic Period, had entered the shoemaker's ken; and today in Europe where custom shoes are made in master's custom shops, the term of apprenticeship is still seven years. In the schools of New England of the eighteenth century the boys were taught to write the form of apprentice indenture papers which were in current use in England,[2] so that both future masters and apprentice lads became familiar with those terms along with their Rule of Three and the injunctions of the Complete Letter Writer.

Phase 2. Extra sale work

No one year can mark the transition from the first to the second phase of the Handicraft Period, the change from the purely bespoke work to the partly sale work, for though these phases were inspired by different ideas, one and the same man as shoemaker made the transition personally without knowing himself that it was a transition. The time was bound to come in the increasingly prosperous Commonwealth of Massachusetts when the numerous trained master shoemakers in any one town would have to compete with each other for customers, or else drift out from towns like Boston and Lynn to get work on the frontier. Interesting stories are told of the way Josiah Field, of Boston, came to Randolph and taught the shoemakers there the first " city ways." Since probably the best shoemakers did not need to drift to the country, the standards which the others, as masters,

[1] A shoemaker tied up his tools in his leather apron. See legend and song, pp. 40–57, History of the Gentlecraft, etc. Quoted in Appendix III.
[2] Cf. Ebenezer Belcher's Exercise Book of 1793. Harvard University Archives.

would pass on to their journeymen and apprentices in turn would be something less exact, and somewhat out of touch with town standards, though the lack of sale shoes then, either in the stores, or on the feet of fellow townsmen, would keep his village customers in happy ignorance of the fact.

The making of unordered shoes for stock gradually became a practice. At first, even in the larger towns like Lynn and Boston, where the best shoemakers were grouped and could secure the pick of journeymen and apprentices, there must have been times when shoes, made for a definite customer as bespoke work under the first phase of the Handicraft Stage, were spoiled by poor work or mismeasurement, or left on the hands of the master workman through caprice or failure to appear on the part of the customer. This was probably the first entering wedge in hundreds of cases for the second phase of the Handicraft Stage, *i. e.*, a time of extra sale work, and it repeated, at an interval of four or five centuries, the history of the shoemaking craft and trade in different parts of Europe. The sale of such a pair of shoes could not fail to suggest a new possibility to custom makers. This new market helped to relieve the fear of a like misfortune in the future, and to give an idea of what they could do in slack times when the apprentices and journeymen might fairly be expected to "eat their own heads off" to the shoemaker's loss while he was waiting for definite orders from specific customers. In such a case, the craftsman [1] ventured to make up stock which he had on hand, thus employing this otherwise wasting labor, and then tried to dispose of the shoes either in his front window; or, lest by doing that he compete with himself when customers came to his shop, he put them in the village grocery store to help out his account, just exactly as the farmer's wife of today puts in some eggs or a few jars of preserves which she ventures for sale in more or less of a gambling spirit.[2] Since the market was uncertain and slow for this extra work, both the stock and labor may frequently have

---

[1] Cf. Hall's Book of the Feet, a History of Boots and Shoes, pp. 81–84, for similar manufacture of sale shoes in England.

[2] Shoemakers were not alone in sending "extra sale work" in the way of shoes to the general store of the village. Account books of such stores show that ginghams

been below the standard used in custom-made shoes. It might naturally happen that the demands would be more steady and the profits could be relatively higher for this lower cost work when once it was introduced into the community, even when it was all done at the direction of the same master shoemaker in the same shop and by the same workers, by simply using different standards and different grades of stock.

With this idea, when once started, of making some *extra sale shoes to fill out time at custom work*, shoemakers were bound to become increasingly pleased, especially those who lived in villages too far from Boston to attract non-resident customers, but near enough to send their surplus product into Boston. A seemingly typical though belated case, with all its local flavor, can be followed in detail in the bills, letters, account books, and oral traditions of Quincy Reed, of Weymouth. He expected to be a shoemaker just as his great grandfather William, who landed in Weymouth in 1635, and his grandfather and father had been. In 1809, the father was a master with custom work, doing some sale work for local consumption. As Quincy tells the story: "My brother Harvey began it by taking chickens to Boston. He had a pair of chaise wheels in the barn, and putting on a top piece, loaded her up and drove to town. He hung some shoes on the chaise and we sold them in Boston. We did not have a wagon then — I can remember when there was n't a wagon in this part of the town, and between here and East Abington there was only one pair of wheels. All the shoes, before we began business, were carried into

and linens and worsteds were left by weavers quite regularly just as butter and cheese were.

The custom of women of the present day, who knit or crochet, or paint dinner cards and valentines as a by-industry and place them for sale in stores organized and maintained for other purposes, thus saving the charge or cost of rent and service as well as securing a likely place in which to expose their wares, is an interesting survival rather than revival of an old custom well known in the later years of the eighteenth century and the first quarter of the nineteenth century when staple wares, like cloth and shoes, were exposed for sale in the same way.

Interesting records of stockings as " sale " goods, and of contemporary custom shoemaking, are contained in the Timothy White papers reprinted in the Nantucket Historical Collection. We seem to have in them the account of a farmer who taught school and made shoes. See. Appendix VII. Cf. Appendix VIII for the George Reed papers.

Boston in saddle bags. We hired a store of Uriah Cotting, at 133 Broad Street, and fitted it up. Then I used to keep a chest of shoes in a cellar near Dock Square and on Wednesday and Saturday would bring out the chest and sell. I got fifteen and twenty dollars a day by it in 1809. I was sixteen and my brother was eighteen years old then. We moved into the Broad Street store with two bushels of shoes. I used to cut out what would promise to be $100 worth a day. We could n't have them made equal to that, but I could cut them. One day I cut 350 pair of boot fronts and tended store besides. Most of the shoes were made by people in South Weymouth. We had nearly every man there working for us before long. Used to bring out the sole leather swung across the horse's back in those days. We did n't have any capital to start with except father's assurance that 'the boys are all right and will pay their debts.' When we got of age Harvey paid father $1000 for his time and I paid him $3000. By then we had got up a stock of $10,000, and I have the inventory now to prove it. We are getting $2 for the best shoes, and $1.25 to $1.50 for the West India shoes. . . ." [1]

These pages from the Reed papers have been quoted in detail because they are important evidence of a typical transition from the second or sale work phase of the Handicraft Period to the opening phase of the new Domestic Period of the shoe industry.[2]

[1] These Reed papers are kept in the old Reed house in South Weymouth, Mass., where Mr. Quincy Reed generously allows students access to them.

[2] Bryant, in his Shoe and Leather Trade for One Hundred Years, mentions retail and jobbing houses in Boston at the close of the eighteenth century giving most conclusive proof that the custom work had been gradually giving place to large amounts of sale work under the Domestic System for export before 1800. To quote his facts in part: "In 1796, Perez Bryant & Company had a shoe store at No. 66 Ann Street, Boston, and a store in Savannah, Georgia, to which it made large shipments. This Perez Bryant was a native, like Levi Leach (see p. 138), of Halifax, Mass. Silas Tarbell also had a store in Ann Street, Boston. By 1798, E. Thayer & Company had a store there in Ann Street and made large shipments to Charleston, S. C. and to Savannah, Ga. Amos Stetson had a large store in Mercantile Row, and made similar shipments. Both Stetson and Thayer were from Randolph, and the former gave a handsome town hall to his native town."

Asa Hammond's store was at 14 Ann Street in 1798. Samuel C. Torrey was a tanner in Pleasant Street, Boston, and had his tan yard there.

These facts would make us feel that Reed was simply one among many followers of the pioneers in the Boston export trade.

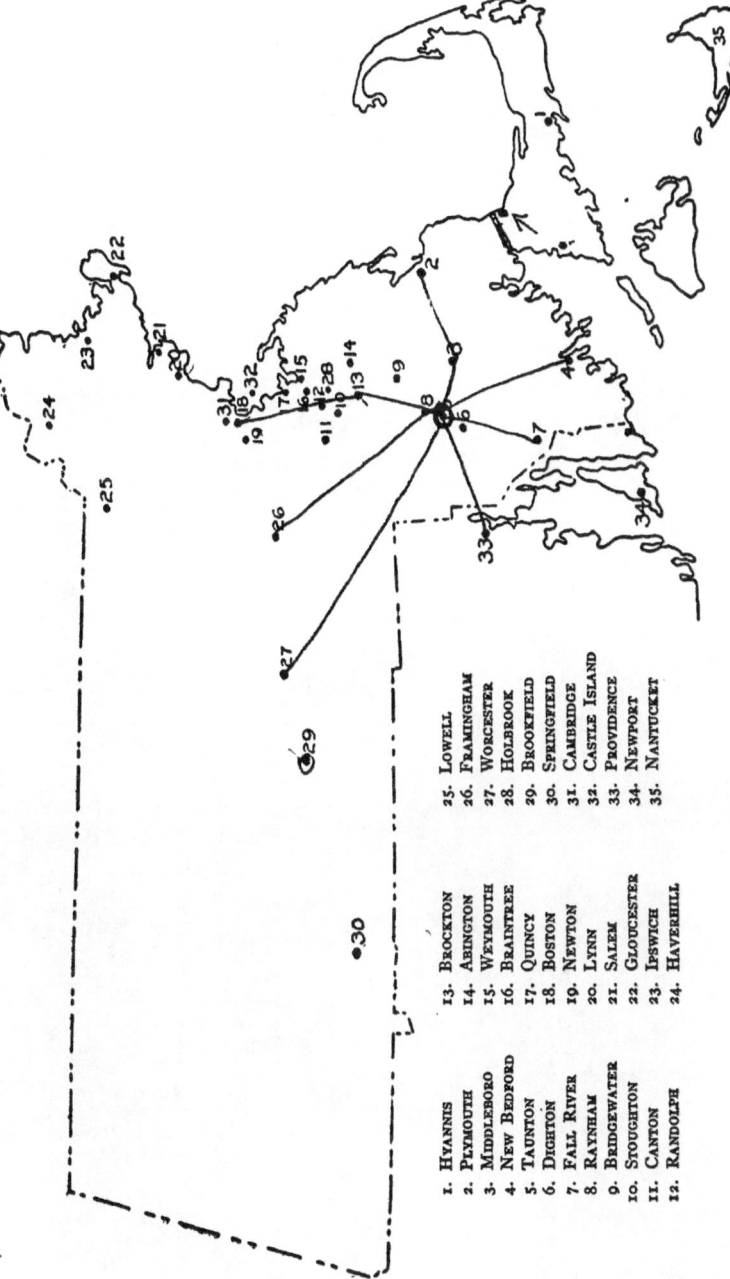

They show how custom work, as the essential feature of the shoe trade, gave way whenever and wherever there was developed an accumulated stock of extra sale shoes, with capital and enterprise enough to make shoemakers [1] turn into producers for unknown customers. In Massachusetts, this new period in the shoe industry was contemporary with the Revolutionary War period.

Roughly speaking, the Home and Handicraft Periods had been contemporary with the settlement and colonial eras in Massachusetts, that is, with frontier conditions which were primitive compared to those which the settlers left behind them in Europe.

BRIEF SURVEY OF THE ECONOMIC CONDITIONS OF EARLY MASSACHUSETTS, IN WHICH THE HOME AND HANDICRAFT STAGES WERE DEVELOPED

One has to recall vividly colonial conditions and get into full sympathy with their demands and limitations, to understand the reversion to these Home Stage productions made by the European colonists who had left in their homes, whether in England, Holland, France or Germany, a well developed organization of industry for all the crafts ranging even in small towns from the saddlers to the silversmiths, from the weavers to the candle makers. There the Handicraft system in its gild form, already merging rapidly into the Domestic System, had taken the place of the Home Stage production some centuries before. Probably few settlers who came to America at any time in the seventeenth or the eighteenth centuries were accustomed to the mediaeval gild system, and certainly few of them knew of the Home Stage unless they came from frontier towns and isolated valleys in Europe, which they scarcely would have left for similar frontier life in America. So most of the colonists took up for the first time the Home Stage production, which is always typical of frontier life and isolated homes, whether in a mediaeval manor, a colonial

---

[1] Quincy Reed must have been cordially disliked and distrusted by custom shoemakers who had none of his ambitions. He approached the industry not as a shoemaker, but as a trader. A custom shoemaker would hate this newcomer as much as he had the outgoing itinerant cobbler — considering both as makers of "bad ware."

farm house or a log hut; typical alike of the absence of stores and cash, of diversified industry and spare money, of division of labor, of means of exchange; in short, typical of industrial conditions in a community that is young. The Home Stage of production went hand in hand with pioneer conditions in Massachusetts and in New England in general.

We need to have a picture that gives the details [1] and settings in the economic and social life of several of these Massachusetts towns in order to bring such conditions before our eyes. Much has been written of the early political and religious life and problems of these same towns, but little about the economic or industrial. Yet between the lines of the documents which have been preserved carefully because they related to political and religious affairs, we can get the story we are seeking.

The commercial life and the industrial life of each community, whether seaport or inland town, whether in full contact with surrounding communities or isolated, always determined the economic life within its bounds. The stages and their phases of industrial organization are independent of time,[2] and of advance in other places. Each community's development has determined not only its local industries, but also the time when each industry should appear and should pass from one phase to another.[3]

A new Massachusetts town in the early colonial days was settled either directly from Europe, or from another colony, or from a parent town. Either the desire for more land or more religious and social freedom led settlers out of the regular settlement into a neighboring fertile valley, or into a tract remote enough to escape the political and religious system of the town, though not of course of the Massachusetts General Court, which made laws alike for every town and village. Its surveillance could not be as detailed or strict in the little new frontier settlements, so that there was greater freedom to go hand in hand with the greater discomforts, because by the remoteness of the settle-

---

[1] Cf. Appendix VIII, IX, X, for such details from the original sources.
[2] See Appendix VIII.
[3] Bücher, Die Entstehung der Volkswirtschaft, chapter IV, in Wickett's English translation of the 3d German edition, pp. 150–184.

ment, the distance " to mill and to meeting " became greater and acted as a barrier to frequent intercourse with the traders and the neighbors in the old town. While for political and religious purposes the colonist went back to the old centre on foot or on horseback, for social and economic life he was forced to depend upon his own family and farm. This condition of things held until there were enough settlers in the valley or on the hilltop to decide in a public meeting to have a new village of their own. The colonists, however, undertook first to be, not towns or villages, but just precincts of the parent town. They clung to the main road from Boston or Plymouth, the kernels of the first two colonies of Massachusetts Bay, even while they pushed their way forward to some neighboring settlement. This is the way that Randolph as a part of Braintree, and Brockton as a part of Bridgewater, originated, and these two towns have been studied in detail along with Lynn and Brookfield, because they have become leading centres of the shoe industry in Massachusetts.[1]

Randolph

From the old town of Braintree, which lay on the Bridgewater Path from Boston to Taunton and Plymouth, there had come settlers enough toward the south to form a village by Cochato River and Tumbling Brook in the first quarter of the eighteenth century.

On details of its growth and its problems such as are given in Appendix IX and taken from parish and town records of Braintree and Randolph we have to depend to understand the economic conditions of a community, which became, before the century was over, one of the leading shoe towns of Massachusetts.

By 1762, either by the sale of surplus produce, or by wage-labor, or by the income of grist and saw mills, the South Precinct of Braintree was earning money as well as making its living. New demands for labor came in with the odd cash to pay for it.

[1] Even at the risk of seeming discursive such details of local history as can supply a background for the understanding of the social and economic conditions of these towns are given in Appendix IX.

There must have been by this time a non-farming class, or at least a non-land-owning class who were artisans. Some of these we know were shoemakers. Before the Revolutionary War, there were numerous pits, and a regular bark house[1] for tanning, at the

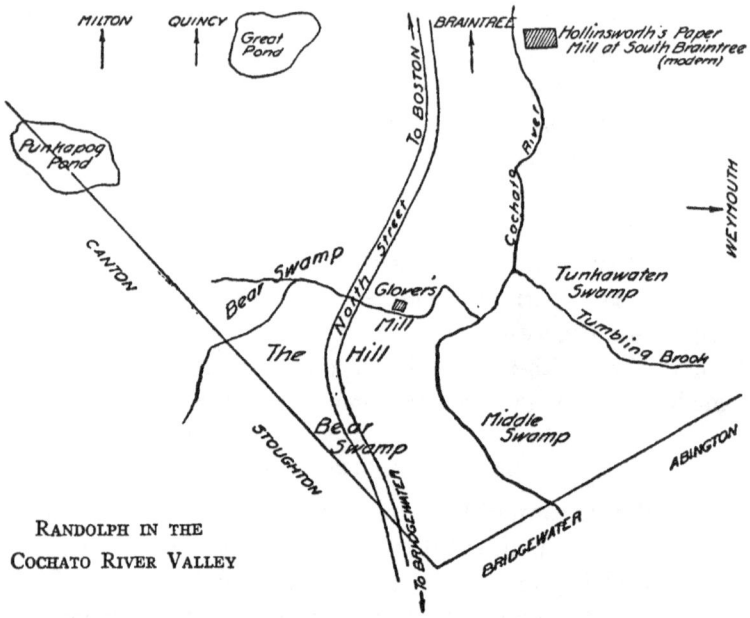

RANDOLPH IN THE
COCHATO RIVER VALLEY

south end of the town on the Bridgewater road, and another at the north end of the town known as Thomas French's vat. The shoes of the South Precinct were made of leather raised, tanned

[1] Bark pits for tanning may be seen on map, p. 176. The evidences of a tannery and bark house on the Wales estate in the south end of the town persisted into the nineteenth century. Built some time before 1770, the bark house gave its name to the "Bark house piece" of land throughout the next century in the Wales family. Old Thomas Wales, deacon of the first Randolph Meeting House in 1731, was the founder of that Wales estate to the south of the Hill. His son, Dr. Ephraim Wales, 1st, who was graduated at Harvard in 1768, had two granddaughters, Annie and Sallie Wales, who were living at this old homestead in 1915 when they gave me these facts about the pits. Their mother, Polly Alden, was the daughter of the Silas Alden, 1st, who will be mentioned again as an early entrepreneur. When Polly Alden married Dr. Ephraim Wales, 2d, and came to live at the old Wales home, the piece of land a little distance back of the house and enclosed by a stone wall was full of pits where leather had been tanned on the "bark house piece," and she used to explain them thus to her daughters Annie and Sallie.

and fashioned on the latter spot by handicraftsmen before the Revolution. This settlement by the Cochato River had passed through the Home Stage of the organization of the shoe industry, and was well advanced into the Handicraft Stage before it became politically independent, or took its share in the Revolutionary War. By curious coincidence, some of its citizens learned much about their trade even while they served in the Continental cause.[1]

On March 9, 1793, Randolph [2] was incorporated as a town just south of Braintree.[3] It was no longer a mere farming community that was self-sufficing; it was interested economically as well as politically in the outside world where it sought markets wherein to sell as well as buy. The Handicraft period of shoemaking closed with the Precinct days. Mr. Silas Alden was manufacturing boots in a shop which stood on his own land, the work being done by apprentices and journeymen, but instead of making merely " bespoke work " for different customers, as he and fellow shoemakers had in the first phase of the Handicraft period, he took the boots and shoes to Boston for sale, carrying them in saddle-bags as he rode on horse back over the Blue Hills through Milton.

Thus by the time Randolph had become incorporated as a town, at the close of the eighteenth century, the Domestic Stage of the shoe industry had appeared there in its first phase.

[1] Details from Kingman, who tells about shoemaking at Castle Island, in Boston Harbor, where Massachusetts Colonial soldiers were stationed with English regulars, will be found in his History of North Bridgewater. See also Appendix XIV.

[2] During the last years of the war there was talk in the village of independence, not only for the United States of America, but for the South Precinct of Braintree as well. By 1792, this parish had formally petitioned both Braintree and the General Court of Massachusetts to be incorporated as a town, and they had chosen to name it after Peyton Randolph, of Virginia, the First President of the Continental Congress of 1774. The petition was signed by one hundred and twenty of the citizens representing a population of at least 600 people, and was accepted.

[3] Braintree was the mother town not only of the Cochato settlement, which became Randolph, but also of a North Precinct, which became Quincy, the home of Presidents John Adams and John Quincy Adams.

BROCKTON

(Formerly North Bridgewater)

Meanwhile the town of Bridgewater in the Plymouth colony, founded in 1656, had been undergoing the same sort of changes as the town of Braintree. As a parent town, it was obliged to allow its growing children to depart in peace, and to give them a portion of its lands. The North Precinct (now Brockton) was situated along the Bridgewater road, north from Bridgewater, and six miles from Randolph, which lay on the direct road from Bridgewater to Boston, through Braintree.[1]

A list of parish rates in 1744 and of polls due in 1770, together with the estimate of population for 1764 give us some definite ideas of the growth of the North Precinct of Bridgewater in its first fifty years. Its citizens took their share of colonial burdens in Indian fighting, not on their own borders in self defense, as in the case of the Brookfield men, but off on the Massachusetts frontiers. We have lists of those serving in the French and Indian War, and those fighting in the Revolution.

To the men busy with getting cannon and shot for supplying the Continental Army, North Bridgewater was known as an iron smelting and casting community.[2] Cabinet making and handle shaping for iron tools were also rivals or complements of farming as an industry in the precinct. Spinning and weaving, tanning and shoemaking, though not industries for outside consumption, went on in every farm house from the infancy of the precinct until it had passed through the years of the Revolution. The tanneries gradually became larger and fewer, and ceased to be mere "neighbor's pits." The custom of making shoes for one's own family was given up. The North Bridgewater shoe industry seems to have been in the Handicraft Stage during or soon after the Revolution, for trained shoemakers were working for customers at the close of the eighteenth century. It would appear that North Bridgewater was later than Randolph in developing the tendency to "sale shoes" which led out of the Handicraft and into the Domestic Stage of the shoe industry. Perhaps the iron

---

[1] See Appendix IX for details. [2] See footnote, p. vi of Preface.

works and woolen mills offered a better chance of profit and absorbed even the small amounts of capital. It may be, however, that the records of the first sale shoes were not kept, yet when we come to details about Micah Faxon,[1] of Bridgewater, who was making sale shoes in 1811, we find that he was regarded by the next generation in his community as a pioneer. Though only six miles away from Randolph, Bridgewater entered the Domestic Stage a score of years later than its neighbor.

BROOKFIELD

Still further west in Massachusetts the town of Quabaug, or Brookfield, was passing through the Home and Handicraft Stages of the boot and shoe industry in its colonial and Revolutionary days. As Randolph was the child of Braintree, so Quabaug seems to have been the child of Ipswich, an eastern seaport of Massachusetts, but the foster child of Springfield, in the Connecticut River valley.[2]

The year 1783 found Brookfield, as Quabaug had been renamed, the third town in age, and the first, as to its wealth and population, in the county of Worcester. It contained four hundred and thirty-eight dwelling houses and three thousand one hundred inhabitants, while its neighbor, Worcester, had only two thousand one hundred people. Even before the Revolution, Brookfield had developed the manufacture of woolens sufficiently

[1] "Mr. Micah Faxon was probably the first person that manufactured shoes for the wholesale trade in the town of North Bridgewater. He came from Randolph in 1811 (was one of the South Randolph — Brookville — Faxon family for whom Ebenezer Belcher worked) and commenced cutting and making shoes in the house that was formerly occupied by the late Matthew Packard, and on the same lot that Mr. Faxon's house now stands. At that time there was no one in town that could bind the vamps and put the shoes together, and they were sent to Randolph to be made. At first he made one hundred pair of fine calf spring-heel shoes and carried them to Boston on horseback. His first lot was sold to Messrs. Monroe and Nash, a firm on Long Wharf, Boston, who were among the first to send goods to the South. When carriages came into common use, he carried his shoes into the city in wagons and brought out his own leather. The market-men and those that carried wood and other goods to market, used to bring out stock for him, which, of course, was in small lots at first." — *Kingman's History of North Bridgewater*, p. 405.

[2] Cf. Appendix IX for details of the early political and economic history of Brookfield.

to make it, like Bridgewater, with its iron works, an illustration of the American manufacturing communities, which the jealous English Parliament tried in vain to suppress. Along two rivers there were enough good water privileges to provide power for owners of saw mills, grist mills, and woolen mills as well.

There is a description of Brookfield, made by Dr. Snell, who became pastor there in 1798. Though his account[1] as here quoted was written over fifty years later, it was accepted then and it is now by other Brookfield historians.

> The inhabitants were all husbandmen. Even the few mechanics who wrought at their trades merely to supply town customers, were farmers upon a larger or smaller scale. There was not more than a single mechanic whose wares were purchased abroad; while we were wholly dependent upon other places for most kinds of mechanical business no less than for merchandise. The population of the precinct of North Brookfield was about eleven hundred, nor did it vary essentially for nearly thirty years. The forge and the mills on Five Mile River were the main business centres, and Salmon Dean had a tannery in Spunky Hollow. Francis Stone had another tannery at Waits Corners, while David Thompson and Daniel Weatherby had a tan yard opposite the East Hill place. The town supported four carpenters, but every thriving man could hew, mortise, and lay shingles. The cobblers of that day were Ezra Richmond, who had a small shop in the east part of the town; Malachi Tower, who lived in the old Dempsy house; Thomas Tucker and Abiel Dean who had benches in their kitchens, but used to go round to the farmers' houses in the fall with their kit and stay a week or so, mending and making the family supply of shoes.

When the boot and shoe industry in Brookfield passed out of the Home Stage suggested by the last facts quoted from Dr. Snell, and into the first phase of the Handicraft Stage, it is not possible to state, yet the transition may have come before the Revolutionary War, and probably before even the advent of the woolen mills in 1768. This first phase of supplying boots made to order for actual customers seems to have lasted until the first decade of the nineteenth century, when Oliver Ward and Ezra Batcheller came from Grafton about 1810 and began to work on sale shoes. With the advent of Oliver Ward, and his manufacture of sale shoes, or at least extra shoes to offer in the local general store, the boot and shoe industry passed through the second phase of the Handicraft Stage, and was on the eve of

[1] Cf. Temple, History of North Brookfield, pp. 266-267.

adopting the Domestic System. Thus Brookfield and North Bridgewater came to the opening of the third or Domestic Stage of their shoe industry at about the same time.

Lynn

Meanwhile Lynn had far outstripped every one of these towns in its rapid development, not only of organizing the shoe industry for production, but also for trade. It had secured markets with English colonies and with England herself before the Revolution, and had passed, as will be seen in the following detailed account, out of the Handicraft Stage into the third stage, the Domestic, far in advance of the other Massachusetts towns.

The development of the shoe industry in Lynn, a seaport located only a few miles from Boston, is typical of that in towns which came in contact with European and other foreign markets. There were several reasons why Lynn should pass out of the Home and Handicraft Stages of the shoe industry before such towns as Randolph, North Bridgewater and Brookfield, but its beginnings were practically the same in economic conditions and problems, except for its closer contact with the coast and coast trade, which lessened its industrial and intellectual isolation.[1]

The arrival of two shoemakers from England in 1635 determined Lynn's chief industry for the future. Attracted by lands or friends or the Ingalls tannery, Philip Kertland and Edmund Bridges,[2] came to Lynn. They either took apprentices and spread the fever or found others there in the farming community all ready to join them. From then until 1750, no one seems to know any more about the shoemaking industry in Lynn or any of the surrounding towns. Probably it passed through the Home Stage more quickly than any other town in Massachusetts, for, by 1650, if not earlier, it had passed into the Handicraft or Custom Stage, out of which in turn it passed before the Revolutionary War.

[1] Cf. Appendix IX for details of the political and economic history of early Lynn.
[2] In spite of the tradition that Edmund Bridges was the second shoemaker in Lynn, he seems to have been a blacksmith rather than a shoemaker by trade. There is a court order of May, 1647, about his neglect in "shoeing Mr. Symonds' horse."

CHAPTER II

DOMESTIC STAGE. PUTTING–OUT SYSTEM, 1760–1855

PHASE 1, 1760–1810

The Domestic Worker makes the complete shoe, but for a market, not a consumer. Export trade opens. The United States tariff protection of shoes begins. Shoe manufacturers are not always shoemakers, but simply entrepreneurs seeking profitable investments.

Phase 1 defined and characterized.

Rise of Phase 1. Newspapers as a source of information on the shoe industry — as to goods, materials and tools, tan yards, imported hides, supply of shoemakers, wholesale and retail trade, shoes for export to the South and West Indies. Private and official correspondence about shoe orders; Wendell letters and accounts; account books of Breed, at Lynn.

Summary of facts about Phase 1.

DOMESTIC STAGE — PUTTING-OUT SYSTEM

Phase 1. Domestic Worker still makes the complete shoe (about 1760–1810)

THE close of the second phase of the Handicraft Stage, with its growing attention to extra or sale work, appears so like the opening phase of the new or Domestic Stage that it is not readily distinguished, especially where the handicraft workman changed naturally and gradually into the domestic worker employed by a shoe merchant, who, as an entrepreneur, marketed the goods. Yet the two stages are fundamentally different. Just as the *Handicraft Stage is characterized by the direct dealing* of the shoemaker with his market, and his dependence upon his own skill and efforts in making as well as selling, so the *Domestic Stage is characterized by the indirect dealing with the market on the part of the shoemaker,* who is simply to manufacture the boots and shoes which a capitalist-entrepreneur markets at his own risk and profit, supplying in whole or in part the tools and materials.

Three phases of the century-long Domestic Stage can be traced, defined, and illustrated. The first, coming gradually by the middle of the eighteenth century, prevailed in Massachusetts by the close of the Revolution and lasted through the opening years of the nineteenth century. Though apprentices and journeymen were employed, the less skilled and more irregular labor of the women and girls of the family was also utilized. The shoemaker turned over to the entrepreneur the completed shoe, often the combined labor of every member of his family besides his apprentices and journeymen, but with all the processes done in his shop under his direction. This last fact, that the *domestic worker still made the complete shoe*, is to be considered a special characteristic of the first phase of the Domestic Stage.

Rise of capital for the shoe industry

The financing of the shoe industry is perhaps just as much of a characteristic of this first phase and continues to be in phases two and three. In short, it is the most vital common factor in the whole Domestic Stage. For the first time capital, in the modern conception, was necessary in the shoe industry. Shoemaking in the Home Stage was done by the shoemaking farmer for his family to avoid spending money, and to avoid going without footwear when no stores or labor were at his command. Then the itinerant cobbler appeared to work for board and lodging with or without a small wage. He had no stock or capital involved save his kit of tools; the farmer-consumer, as before, provided the stock.

Shoemaking in the Handicraft Stage in its first phase of ordered or bespoke work was purely a matter of a shoemaker's earning a living, not of seeking and making investments. Even in its second phase of sale work, it was largely a question of saving from waste the spare time of journeymen and apprentices, of poor stock, and left-over pieces or of misfit stock left on hand, that induced shoemakers to make sale shoes. This sale work, coupled as it often was with the expansion from a town to a national economy, with the changed conditions from provincial to national political and commercial life, was the first taste of venture, or risk, or profits on investments which led over into, and made possible, the Domestic Stage in the shoe industry. It was openly one of capitalism whether the entrepreneur was a shoemaker "born and bred" or whether he was an outsider attracted by the chance of profit which the export trade offered to the shoe manufacturer.

The Domestic System and the Rise of National Markets for the English Colonies in America

Similar conditions and developments can be traced in the trades in England. The woolen industry and trade there, for one illustration, presents decidedly interesting parallels to the shoe and leather industry from 1760 to 1855, and even to 1875 in this country. At first and until the middle of the fourteenth century, all the different branches of labor requisite for turning out a proper finished piece of cloth were carried on as separate industries by independent workmen with apprentices and journeymen in their houses. The weaver bought the wool or yarn and made cloth. He sold the cloth to a fuller who made it into a close fabric. The fuller then sold it to a shearman who smoothed the nap with his heavy shears and sold it to the purchaser. In the middle of the fourteenth century, a new class of drapers appeared, who seem to have been merchants. They bought cloth of weavers or of fullers and sold it to customers in distant markets. This was when the considerable export trade in cloth began. Later the clothier delivered wool to the weaver, who employed carders, spinners, dyers, fullers, and other workmen. The capitalist-clothier furnished the materials, arranged for the various processes, and sold the finished product. The weaver might actually do his work at home (domestic worker), but so far as economic relations were concerned, he was dependent for stock upon an employer (entrepreneur) who later on furnished the implements also.

Unwin[1] gives a concise statement of the appearance of the domestic system in various industries in England, which is apropos for students of its advent in the shoe industry in the late eighteenth and early nineteenth century at the time of our Revolutionary War, the Continental Wars, which affected our foreign

[1] Industrial Organization in the sixteenth and seventeenth centuries, by George Unwin. Published at Clarendon Press, Oxford, 1904. Chapter I.

commerce and markets, and the War of 1812. Unwin, speaking of the advent of national markets which made the more important industries freer to concentrate in favorable localities and lose their local limitations, said:

> The craftsman might continue, as the currier did, to work in his own home, using in part his own capital, although dependent for constant employment upon the larger capital of others.[1] On the other hand, he might become a mere wage-earner in the workshop of a capitalist master, who combined several crafts under one direction. The domestic system was the result of the adoption of the first of these alternatives. The second contained the germs of the factory system. Both forms of change were resisted by the craftsmen whose independence was threatened, but whilst the opposition to the second was backed by a strong public opinion embodied in persistent legislation,[2] the first proved, in the case of the more important industries, to be an inevitable necessity of progress.

In the American colonies, there was no legislative struggle. The transition was allowed and oftentimes not even sensed, since in our elastic and growing market, changes were easily made and not always recognized as important. This progress was strongly desired by the new capitalist-merchant class of shoe manufacturers, and little feared by the shoemakers. While in England the gild system had emphasized the vested interests of the craftsmen, in the American colonies, shoemakers did not expect nor fear the loss of their economic independence.[3] The shoe industry in this country, therefore, escaped struggle and legislation to

---

[1] Ebenezer Belcher and Samuel Ludden are illustrations of just this condition. Cf. pp. 46, 48.

[2] Unwin gives the following as footnote to the above mentioned legislation: "The English statutes relating to the leather industries afford the most striking illustration of this. The Act of 1389 forbade tanners to be shoemakers or shoemakers tanners. This was renewed in 1397 but suspended in 1402. In 1485 the tanners were forbidden to curry and curriers to tan. In 1503-04 the curriers and cordwainers were prohibited from interfering with each other's trade. Under Elizabeth and James I, the limits of each trade were marked more precisely and its technical operations minutely regulated. But these laws were found so irksome that Elizabeth empowered a favorite by letters patent to grant exemptions; and in 1616 the London Cordwainers and Curriers after much litigation had come to a mutual tacit agreement to ignore them." Unwin: Industrial Organization, pp. 21-22.

[3] Perhaps the journeymen shoemakers of Philadelphia were exceptions. See Commons in Q. J. E., November, 1909.

prevent the capitalizing of it, and the inevitable accompanying attempt at concentration of allied or supplementary industries. From advertisements we shall show tanners and curriers and shoemakers on friendly terms and the entrepreneurs controlling alike the stock and work of them all. Correspondence and account books will show merchants seeing to it that enough leather was tanned and brought to market to ensure proper supplies for boot and shoe making for their rapidly increasing export possibilities.

At the outset of the Domestic Stage in Massachusetts (in some towns as early as 1760) the market was widening and the possibilities of profit were as good in the shoe trade as in other ventures in the East and West Indian trade. The development of the retail and wholesale trade in cities like Boston and Philadelphia tempted capital. There were already more shoemakers than could be supported by custom work or by the small ventures of extra sale work. Both the workers and capital were ready for ventures on a larger scale for both the domestic and foreign markets. Even if foreign-made shoes were being imported, the Lynn shoemakers were undaunted. They studied the materials and the make of the imported shoes while newspaper readers studied such advertisements as the following:

Women's brocaded silk shoes, Women's and Children's Calliminco and Morocco ditto, Galoshes . . . at the store of Wm. Merchant opposite the Golden Ball on the Town Dock, Boston. — *Boston Gazette*, April 23, 1754.

Samuel Abbot has imported from London brocade, russett shoes, plain shoes, silk cloggs, soles for men's shoes. — *Boston Gazette*, January 7, 1755.

Influence of Foreign Shoemaking Standards

Foreign stock and styles were bought and duplicated by the Lynn entrepreneurs. They used the imported materials for uppers and findings. Boston merchants [1] supplied Lynn shoemakers with camblets, everlasting, callimancoes, awl-blades, gimlets, fourteen ounce tacks, shoe nails and hammers.

The next step for the Lynn shoemakers was to depend upon an "imported shoemaker" to teach them additional niceties of the

---

[1] Amos Breed's account books show him to be a large consumer for all such materials. Cf. Appendix XI.

craft. John Adam Dagyr, a Welshman, came to Lynn in 1750 and "gave great impulse and notoriety to the business by producing shoes equal to the best made in England."[1] Whether he came on his own account or was invited or even hired to come by some entrepreneur shoemaker, is not known. His coming at just the right moment may have been, and probably was, only a very fortunate coincidence in the history of Lynn shoemaking. Ten years after the *Boston Gazette* advertised those competing English shoes in 1754, it printed in its issue of October 21, 1764, the information:

> It is certain that women's shoes made at Lynn do now exceed those usually imported, in strength and beauty but not in price. Surely then it is expected that the public spirited ladies of the town and province will turn their immediate attention to this branch of manufacture.

Influence of Revolutionary and Patriotic Spirit on the Shoe Industry

Even then, the advertiser made use of the spirit and temper of the times. Non-importation sentiment was gaining force, people were to be provided with chances to prove their avowals of belief in patronizing home manufactures. Evidently patriotism plus shipping conditions made the proper response come in the next twenty years. In the single year of 1768, there were 80,000 pairs of shoes made in Lynn according to the statement which appeared in the *Palladium* for February 6, 1827.

Newspapers as a Source of Information on the Shoe Industry

A study of the newspaper advertisements,[2] of account books and of private correspondence from about 1750 to 1810 makes a definite addition to the information obtainable concerning the rising wholesale and retail trade in boots and shoes, the prices, and the volume of business.[3]

---

[1] Alonzo Lewis in his History of Lynn, published 1844, p. 91, makes the claim, quoted above, basing it upon traditions and earlier printed accounts.

[2] See Appendix XII and XIII for samples of such advertisements.

[3] The London archives hold records for 1771 showing the shipment of 5938 pairs from North America to the British and foreign West Indies. Of these, 1500 pairs of shoes were from Philadelphia. The bulk of the shipment probably came from the

Importations of Foreign Shoes and Public Vendues of Sale Shoes

Some of the advertisements [1] show the competition from foreign imported shoes, especially women's shoes, which had to be met by Lynn. Others show conditions in Portsmouth in the neighboring state of New Hampshire, where public vendues of shoes were announced, and other features of the industry as organized in Massachusetts were being repeated. Some "stop thief" notices show that not all cobblers were honest.

During the five years immediately preceding the Revolutionary War, there were naturally fewer advertisements of English-made shoes, but shoemakers' supplies of goods and tools were still imported and advertised. Once in January, 1771, the *Massachusetts Spy* advertised a cargo from Leith, which included one cask of shoes. Here, as in our later trade with the West Indies and the South, the watertight container was in constant use for packing boots and shoes as well as other commodities for export. In the same paper, under date of March 11, 1773, shoe blacking appears for the first time to my notice, offered as "patent cakes for making the liquid, shining, blacking for shoes."

Sole Leather, Imported and Domestic

Boston merchants were importing dried ox hides from Jamaica by February 21, 1774, according to the columns of the *Boston Gazette*, and the tanner's business was thus taking on competitors from the West Indies as well as from England. By 1787, the Salem merchants [2] were importing the "best of salted Hides lately from the Cape of Good Hope to be sold at the store of Elias Hasket Derby." By 1795,[3] they were importing in one shipment 800 dried hides from Curaçao, which they offered for sale

---

other shoe centre, Boston, and its neighbors. The historical account of shoemaking in the United States Census of 1900 (Part III, vol. ix, p. 754) says that in 1778, men's shoes were made in Reading, Braintree, and other towns of the Old Colony for the wholesale trade.

[1] See Appendix XII.
[2] Cf. *Salem Mercury* of May 15, 1787.
[3] Cf. *The Salem Gazette*, April 7, 1795, and November 24, 1795.

along with domestic hides. These came as return cargo probably from some of the voyages to South America.

Sole leather, in whole sides or cut up into "butt soles," was frequently mentioned in advertisements of imported goods from England and there are interesting evidences to examine in other newspapers of the supply of domestic hides.[1]

The shoe industry in central, northern, and western Massachusetts was developing the Domestic Stage tardily but surely in the decade following the Revolution, taking its share alongside Lynn in export trade before the century was over. Notwithstanding the attention called to the sale of imported leather, a large percentage of the leather made up into boots and shoes all through the eighteenth century was domestic. The village tanneries were numerous, busy and oftentimes saleable.[2]

Barter and Financial Conditions in the Shoe Trade

The lack of ready money, and expected difficulties of effecting future purchases, are much more apparent in newspaper advertisements the farther the place is from Boston and the coast trading towns. The farmers and merchant shoemakers had pretty nearly barter conditions in Hampshire County.[3]

Some farmers used the flax which they were raising to meet a demand of the shoemakers for shoe thread. The *Hampshire Gazette* of December 30, 1789, printed a fairly common advertisement:

> Six Pence per Pound, Will be given for good flax for which pay will be made in Shoes, Boots, Indigo, Tobacco, etc.

Here again is evidence that even a matter of pence in payment was subject to the barter system in western Massachusetts in the troublesome monetary disturbances of the years after the Revolutionary War. James Byers and Co. advertised for sale at their furnace and store in Springfield on December 3, 1788, vari-

---

[1] An issue of November 26, 1770 gives the first advertisement which was devoted wholly to such stock: "Choice Sole-leather. Tanned with Hemlock to be sold cheap for cash by the Quantity or single side. By Peter Boyer and Co. Merchants Row."

[2] See Appendix XIII.

[3] See Appendix XIII, Case of Hezekiah Hutchins.

ous tools and utensils "for which Pearl and Potashes . . . Beef, Pork, Woolen and Linen Check, Country made Hats and Shoes, Old Pewter and West India Goods will be received in payment."

Asa White, of Williamsburgh, in June 1789, informed his "friends and customers" that he had "received an assortment of English and West India goods" which he was now "selling on very reasonable terms for cash and many kinds of Country Produce." He also wanted "a quantity of tow cloth and Butter" for which he would pay "any of the above mentioned goods." Robert Breck, of Northampton, on September 30th of the same year, advertised various goods, including women's shoes, for which he would take home products of farms or industries in return.[1] Western Massachusetts was still a farming section with enough artisans in the town to supply local customers and to make some sale shoes for domestic and export demands.

Supply of Shoemakers

Some newspaper notices [2] give interesting suggestions whence, geographically and socially, the shoemaking class was recruited. We shall see from various journals and account books that the farmers' sons in Massachusetts frequently became apprentice lads, belonging to the same class and town as their master shoemaker. About Philadelphia, however, there seems to have been a steady recruiting of cordwainers from the colored race as well as from Germans and Englishmen who came to this country already acquainted with the craft. At Litchfield, Connecticut, in the same year (1755), the *Monitor* advertised as follows:

> Wanted as apprentice to the shoemaking and tanning business, a lad about 14 or 15 years old that can come well recommended. John Hardy.

This is particularly interesting, for it is very rarely that one finds in the newspaper files of Massachusetts for the last half of

---

[1] These advertisements seem further evidence that many farming sections had passed out of the Household Stage of producing shoes for their own consumption by 1789.

[2] See Appendix XII.

the eighteenth century any advertisements for either apprentices or journeymen, while in the Philadelphia [1] papers for the same time they are common.

Wholesale and Retail Trade

Newspaper advertisements tell us also of the fact that the local country dealers claimed as low prices as could be secured in cities. John Mansfield in the *Worcester Gazette*, of March 6, 1798, offered

a good assortment of shoes, consisting of fancy shoes, black Florentine, misses and children's shoes, all which are made and warranted at his shop, opposite the Bridge and Wheeler's store. Good allowance made to those that *purchase by quantity* and as cheap by retail as they are in Boston. He has also for sale Florentines, Morocco skins, Bindings and Bobbins.

Up in Portland, Maine, in the July 20, 1795 issue of the *Eastern Herald* warranted shoes of all kinds are offered for, "Ladies, Gentlemen, and Children: also Ladies' Sandals and Misses' red and black Morocco Slippers manufactured particularly for Wait and Gedege and sold wholesale and retail at their bookstore [2] in Portland." "Shopkeepers in town and country may be supplied as cheap as they can purchase in Boston."

---

[1] *Pennsylvania Gazette* for January 24, 1771. "Apprentice to be sold at a Reasonable Price. The time of an apprentice lad who is bound to learn the Shoemaker's Trade; he has yet 7 years and 10 months to serve, is a smart likely Boy and is sold for no fault. Any Master Shoemaker who wants such a one, may know the particulars by enquiring. He is living in Germantown on Mr. Wistar's place. Peter Fritsly." Whether this apprentice lad was colored or not cannot be told by the text. There is no reason to think so except in the line where it said "to be sold" and probably here only his time, indentured time, is meant. Another advertisement of February 21, 1771, dated at Lancaster, Pennsylvania, reads:— "Committed to my custody, September 25, 1770, a certain James Hogan 23 years of age 5 ft. 9 inches high, a shoemaker by trade; he says he belongs to Richard Long in Anne Arundel County and province of Maryland. His said master is required to come in 3 weeks and pay charges or he will be sold out for the same. G. E., gaoler."

[2] It may seem odd to find shoes offered for sale at a bookstore, but advertisements show that tanners used to sell supplies of leather to bookbinders and it was not a far cry to cordwainers using a bookbinder's book store for the display of their extra sale shoes, nor for such a bookseller to become an entrepreneur in a shoe trade in the early days of the Domestic System when the venture was simply a "side-show" to his main business.

Shoes for Export to the South and to the West Indies

Manufacture, in other places than Lynn, for export trade makes itself evident through newspaper columns at the close of the eighteenth century. John Mansfield, advertising in the *Worcester Gazette* for March 6, 1799, perhaps suggested wholesale trade for export when he stated that he had for sale

a quantity of excellent Red Morocco skins, Ladies' Morocco Shoes from 5/8 to 8s per pair, — Ladies' Florentine do. from 3/8 to 6s. Misses and Children's Shoes of all sizes.

Hezekiah Beardslee and Co. announced in the *Middlesex Gazette* of February 7, 1795, that

The subscribers want to get made a quantity of shoes of all kinds for exportation for which good pay will be made. Likewise, wanted, good journeymen at the Men's or Women's Branch, none but good workmen need apply — all such will receive good wages at the shoe Manufactory in Middletown.[1]

On June 27th of the same year, this firm advertises for stock.

Wanted at the shoe manufactory in Middletown, Tanned Calf-skins for which cash will be given. Likewise we give 9 pence per pound for green calf-skin.

Salem was even then devoting herself to shipping rather than manufacturing,[2] building up a coastwise as well as an Eastern trade. She was offering [3] chances of coastwise transportation

For Norfolk and Baltimore. The Schooner, Industry, Richard Smith, Master, will sail in 12 days. For freight or passage apply to master on board at Union Wharf.

By studying the newspapers of the South for the years 1780 to 1810, one gets a realization of the extent and method of the

[1] Middletown, in Essex County, was 20 miles north of Boston.
[2] There was some manufacture of shoes there, however, or in outlying towns, or else an entrepreneur bought up stock from a distance, for Abner Chase advertised on December 29, 1794, at Salem: — "Abner Chase has for sale, by wholesale or retail, at his shoe-store in Essex Street, in addition to his former assortment, a variety of thick and thin men's shoes, suitable for the West India or Southern Market. He has also for sale by the piece, a good assortment of Florentine and Sattinetts, cheap for cash."
Salem was then a depot for imported materials as stock for shoes as well as for manufactured exports.
[3] Cf. *Salem Gazette* of January 13, 1795.

wholesale shoe trade for export from New England. In the *Charleston Gazette*,[1] the issue of July 3, 1783, William Hort and Co. advertised that they had entered the commission business, wanted foreign cargoes and country produce, having their counting house on Eliot Street and the Bay. They had just received for sale on commission, in the brigantine St. Peter, from Amsterdam, "textiles, clothing, hats, men's, women's and children's shoes, slippers, wines, cordage, and paint."

By the *South Carolina Gazette* of November 1, 1784, Patrick Hinds advertised to inform factors [2] of Charleston that they might be supplied, on their own accounts with any quantity of Negro shoes and boots for their country customers, — credit until January next. Again at Charleston in the *City Gazette* of May 2, 1790, there was published this notice, —

> Just arrived, by Captain Garman from Philadelphia, and for sale at Harrison's boot and shoe manufactory No. 25 Broad Street, a large and general assortment of shoes and boots made at their manufactory in Philadelphia. ... J. and T. Harrison take this method of returning their sincere thanks to the public in general and their customers in particular for their generous encouragement.

Seven months later, John Harrison advertised:

> Just received from the Northward ... a few hundred pairs of men's shoes and boots and women's stuff shoes made by the best workmen in Dublin. A quantity of brown negro shoes at 3s. 4d. per pair and black at 4s. and 4s and 8d. Gentlemen who give preference to their shoes and boots being made in Philadelphia, their measure will be taken and attended to. Also, received from his manufactory in Rhode Island a quantity of Men's, Youths and Children's fine and coarse shoes. He has on hand to dispose of very low, a number of shoes that came from Elizabeth Town.

These same Charleston shoe merchants, not content with their allied manufactory in Philadelphia, had established one in Rhode Island.[3] The advertisement in the Charleston, S. C., *City Gazette*

---

[1] Charleston, S. C., had a population of 16,000 in 1790.

[2] These factors were general store merchants who supplied plantation owners with everything they needed to buy, from needles and shoes to coffins, and took the planters' crops of rice and indigo in return.

[3] I have not been able to find any clue to the location of this factory, though probably it was in the northwestern corner near the Worcester region of Massachusetts. The shoe business was rarely found in Rhode Island developed beyond the custom stage with some extra sale work. When it came to a consideration of

36   ORGANIZATION OF BOOT AND SHOE INDUSTRY

for July 7, 1791, practically repeated previous advertisements of the Harrison firm and added,

> John Harrison having established a manufactory in Rhode Island[1] for the purpose of having general assortment of low priced shoes, informs the public that he has received from the above place some hundreds of men and children's common shoes, which he will sell at a reduced price.

Private Correspondence of Shoe Merchants

With this information from the newspapers of the last half of the eighteenth century to give a conception of the volume and markets of the shoe industry for both domestic and export trade, one can turn for intimate detail to the private correspondence and account books of shoemakers and merchants.

Captain Thomas Willson, writing to Mr. Thomas Allen, merchant at New London,[2] on January 14, 1773 from St. Croix, said:

> Dear Friend: This is to advise you that I have just finisht the sale of my Cargo am abt. to Sel my Vessel if I can. I have clear'd on my Cargo and Fret £601 ᵇ4 ᵇ4 besides about 49p^r mens Shoes yet to sell a few oats which is not so bad for 1 mo Sarvis. . . .

From the correspondence of Nicholas Brown and Co. of Providence, R. I., under date of 1st month, 4th, 1773, there comes a letter from Lynn, Mass., from Silvanus Hussey, asking Nicholas Brown and Co. "to deliver in Lynn or Boston 100 lbs. tea for 100 prs. women's shoes."[3]

On October 10, 1789, Sir John Temple, bart., H. B. M. consul

putting capital into shoemaking for export trade, entrepreneurs in Rhode Island found that they could do better to engage in shipping to the West Indies as well as to the Far East; in iron and cordage making, and ship building. This was true at Providence, Newport, Warren, and Bristol.

[1] This may simply mean the entrepreneur relations of Harrison with scattered domestic shoemakers in Rhode Island. Cf. the expression "Hamburg embroideries" made in Switzerland but shipped from Hamburg; or "Leghorn hats" from Tuscany.

[2] See letter in MS. in the Allen papers at the American Antiquarian Society, at Worcester, Mass.

[3] Mr. Wm. B. Weeden, author of the Economic History of New England, in transmitting this letter to me added — "tea not specified but without doubt it was Bohea. . . . This letter indicates that manufacturing for trade was then underway. Hussey was a trader probably."

DOMESTIC STAGE — PUTTING-OUT SYSTEM 37

at New York City, wrote thence to the Duke of Leeds, Secretary of State for Foreign affairs, as follows:[1]

> In my last I might have said much more upon the Progress of various manufactures in these States. All kinds of Cabinet Work, carriages, Sadlery, Men's and Women's Shoes, etc. are so made here as that the duties[2] laid upon such articles from Europe will put an End to any Importation of them.

While in most cases the reasons for non-shoemakers becoming entrepreneurs in the shoe manufacture and trade have been left to surmise, certain letters and business papers of John Wendell, of Portsmouth, New Hampshire, tell a story which might be supposed to be typical of an entrepreneur who was pushed, so to speak, into the shoe business as an outcome of other financial cares and responsibilities [3] in settling the estate of a debtor.

Account books kept in Pioneer Days of the Domestic System in the Shoe Industry

The small beginnings in the pioneering days are disclosed in such accounts as those Amos Breed of Lynn kept from 1763 to 1790. In those first years even, much of Breed's work was under the Domestic Stage of the shoe industry.

Cimbil to Amos Breed, Dr.

1763								£	s.	d.
May 7	To make	20	pairs of Callimanco shoes					26	0	0
May 23	"	"	1	"	"	"	"	1	6	0
June 11	"	"	2	"	"	"	"	2	12	0
June 25	"	"	5	"	"	"	"	6	10	0
July 16	"	"	8	"	"	"	"	10	8	0
Aug. 16	"	"	5	"	"	"	"	6	10	0

Brigham to Amos Breed, Dr.

1764								£	s.	d.
Nov. 2	To make	21	pairs of Calla shoes					3	17	6
Nov. 17	"	"	10	"	"	"	"	1	16	8
Nov. 28	"	"	1	"	"	"	"	0	3	8
Dec. 8	"	"	5	"	"	"	"	0	18	4
Dec. 21	"	"	8	"	"	"	"	1	9	4

[1] This letter is in the volume called B. T. 6/21 (*i.e.*, the 21st volume of the Miscellanea of the Board of Trade) at the Public Record Office, London.

[2] Cf. United States Tariff on boots and shoes quoted on p. 40.

[3] See Appendix XV for Wendell letters, especially one of December 23, 1802.

These men, Cimbil and Brigham, are hired to make the shoes. Amos Breed is evidently supplying the stock. He is also keeping a sort of custom-retail trade at his home in Lynn.

Nathaniel Homes to Amos Breed, Dr.

1766							£	s.	d.
Mar. 8	To	1	pair of	Call. shoes for Poeb...............			2	30	0
Apr. 19	"	1	"	everlasting shoes for Hannah.........			2	40	0
Apr. 19	"	1	"	leather shoes for Lydia.............			1	6	0
July 12	"	1	"	silk	"	" self.................	2	8	0
July 19	"	1	"	russet	"	" Poeb...............	2	60	0

The same account with Nathaniel Homes continues for the years 1766–71, with the entry of 52 pairs of shoes in all. Never more than 3 pairs were charged at a time, and generally each pair was for a specified person in the family. All this was custom work, calling for better stock, greater care and skill than Breed and his journeymen were putting into the shoes mentioned above for export.

Breed was doing custom work for Nathaniel Heath, also in 1767–69 for he records him as debtor:

							£	s.	d.
Oct. 30	To	2 pair of shoes, 1 call., 1 leather............					3	10	0
Nov. 14	"	2	"	"	"	leather....................	2	14	0
Dec. 19	"	1	"	"	"	"	1	10	0
1768									
Jan. 7	"	1	"	"		Galoshes......................	1	15	0
Jan. 23	"	1	"	"		calla Galoshes [1].................	2	0	0
Feb. 10	"	2	"	"		leather Galoshes................	2	16	0

Breed not only hired shoes made and sold them for both domestic and foreign trade, but he mended them.

Account of Richard Breed to Amos Breed

1768							£	s.	d.
Jan. 23	To mending 1 pair of shoes							15	0
June 17	"	making	1	"	"	" for wife...........	2	0	0
June 18	"	mending	1	"	"	" self............		12	0
Dec. 15	"	making	1	"	"	"	2	0	0
Feb. 21	"	"		1 Deer Skin....................			7	18	0

[1] Calimanco shoes almost always billed as £2 in custom accounts.

DOMESTIC STAGE — PUTTING-OUT SYSTEM

Meanwhile, Amos Breed was charging the following items [1] to Samuel Breed, who evidently bought things from Amos Breed's general store in Lynn:

```
1768
July 12    To 1 pr. of stuff for self ..............  1   7  0
            "  1  "   "   "    ....................      9  0
            " tools, pinchers and Nipers ..........  1   0  0
            " 2 pr. of heel taps ..................  0   2  0
Aug.        " 1  "  " Everlasting Stuff ...........  1   4  0
Sept. 3     " 1  "  " leather .....................  0  15  0
Sept. 17    " 1  "  " Everlasting Stuff ...........  1   4  0
Sept. 23    " cash paid 25 shillins ...............  1   5  0
Oct. 13     " 1 doz. of Buttens ...................      3  0
Nov. 5      " 1 pr. of Silver Buckels .............  5   1  0
Nov. 5      " 1  "  " stuffs for Gurl .............  0   9  0
```

This account runs thus to 1775. In 1783, October 20th, it is renewed and settled. "The balance due to me (Amos Breed) old Tenor 3-15-11 Samuel Breed. He paid then in full."

These accounts [2] have been given here because they are typical in regard to custom work, sale work for both domestic and foreign trade, dealings with customers and journeymen. To summarize then, briefly, what these newspaper advertisements, public and private correspondence and account books tell us, the demand for Lynn shoes and the competition which makers of women's shoes in Massachusetts had to meet from imported goods had made itself felt by 1760. The Lynn entrepreneurs [3] were specializing in women's and children's shoes and leaving the making of men's boots and shoes almost wholly to custom work or to cruder local work. Few if any advertisements of men's shoes appear before the Revolution.

Tariff Legislation of 1789

During the years of the Revolutionary War, however, the supply of shoes for the Continental army had to be undertaken, and all the stock raised and tanned in this country. Accounts show that organization had developed in the shoe industry to

---

[1] These are selected from many others.
[2] Several others from the Breed papers are given in Appendix XI.
[3] Cf. U. S. Census of 1860, p. lxix.

meet these conditions by the close of the war. Southern and western Massachusetts had come to specialize in men's shoes, brogans and high boots, while Lynn kept to her older specialty of women's and children's. That the boot and shoe business was considered profitable to capitalists and important to the people at large is indicated by the Federal tariff legislation of 1789,[1] which gave protection to this industry. The tradition seems as well founded as it is detailed that this action was due to the shoe merchants of Massachusetts[2] as well as to those of Pennsylvania. Their pressure was strong enough, joined to the general current towards protection, to override the specific illustration of James Madison in a discussion April 9, 1789, against protecting shoes.[3]

Summary of Development in First Phase of Domestic Period

The wording of advertisements all through the years 1760 to 1810 shows us a transition period. Then there were existing side by side both the first phase of the Domestic Stage, where the entrepreneur was a capitalist shoemaker, hiring workers in their homes to make boots and shoes for him to sell at retail or wholesale, and the last phase of the Handicraft Stage of the shoe industry, where the custom makers in the village put some extra or sale work into their local store to help pay their grocery bills.

[1] Tariff rates of July 4, 1789:
On boots per pair, 50 cents.
On shoes, slippers or galoshes made of leather per pair, 7 cents.
On shoes or slippers made of silk or stuff, 10 cents per pair. Tariff of August 10, 1790: provisions of previous tariff kept and one added which put on leather tanned or tawed, and on manufactures of which leather was the chief article, the 7½ per cent ad valorem duty that was put on other raw products.

[2] Bryant's naive account of this legislation influenced by President George Washington's favorable impression of the city is open to question. Cf. pp. 6–7 of Shoe and Leather Trade, etc.

[3] " . . . For example, we should find no advantage in saying that every man should be obliged to furnish himself by his own labor with those accommodations which depend upon the mechanic's art, instead of employing his neighbor, who could do it for him on better terms. It would be no advantage for a shoemaker to make his own clothes in order to save the expense of the tailor's bill, nor of the tailor to make his own shoes to save the expense of procuring them from the shoemaker. . . . The same argument holds good between nation and nation." Cf. Annals of Congress, I Cong. 1834, I, pp. 109–148.

This picture based on printed contemporary evidence is a true one of any community where the close of one phase of an industrial organization runs parallel with the opening of a new phase. In Massachusetts the Domestic System was well established in its first phase for the capitalist-merchant had appeared to venture, to lose or to profit in the boot and shoe industry in some communities by 1760, in many by 1810. The impulse (1) of the sales during the Revolutionary War, (2) of the demands of trade in the United States after it, and (3) of the tariff, gave a big stimulus.

The growth of the market for boots and shoes and its assured protection did not fail to widen the ranks of capitalists and to intensify the manner of production. A new phase, outwardly marked by prosperity and volume, prevailed in Massachusetts from about 1810 to 1837.

CHAPTER III

DOMESTIC STAGE. PUTTING-OUT SYSTEM, 1760-1855

PHASE 2, 1810-1837

Specialization in processes and the rise of the Central Shop are the chief characteristics.

Introduction and definition of Phase 2, "ten-footers," cabbage stock and waste; the central shop and its function; proportion of domestic workers; phraseology changes to keep in touch with new organization; Ebenezer Belcher's account book shows prices and volume of output; Lincoln's store and trade in shoes.

Central Shops and entrepreneurs. Lincoln was an entrepreneur on a larger scale; other investments of shoe entrepreneurs; the Littlefields with the southern markets and the post office; domestic workers and the central shop relations; Natick's beginnings; Felches and Walcotts; Brookfield slow in developing domestic stage of shoe industry; general conditions portrayed in general store account books of Skinner and Ward; F. and E. Batcheller manufactured russet brogans for Southern trade, and built up big prosperous business from 1825 to 1835; general prosperous conditions in the shoe towns in Massachusetts in 1837. Panic and hard times in shoe towns, as well as elsewhere, 1837-39.

Specialization in Processes

THE new phase, which prevailed from about 1810 to the panic year of 1837, was characterized by specialization in processes and the rise of the central shop. The extra capital which was tempted into the boot and shoe industry brought more competition for orders, and suggested a specialization to secure rapid work which may be taken as the chief characteristic of the second phase of the Domestic Stage. The time factor seemed more vital than the quality, for with such big, insistent markets, a merchant could afford to lose a disgruntled customer occasionally, sure of having plenty of others. Already the producer did not have to face the customer. His reputation did not suffer from occasional bad work. The standards were therefore lowered, and the competition of the employers gave entrance into the trade to less skilled and almost unskilled labor to do the cheaper work on men's brogans and women's and children's shoes for the well developed trade in the West Indies and the South. A wide difference arose not only between the quality of custom and of domestic work, but between the wages of "real journeymen" and of shoemakers.

Waste in the Ten-Footers

As the workers grew more numerous in the employ of each Massachusetts capitalist and were scattered in their shops, called "ten-footers,"[1] over a wider area, in fact all through New England from Portland, Maine, to the Island of Nantucket, the need of inspection of the finished product, as well as of saving by the cutting of stock at the central shops became vitally necessary. Previously the stock for both uppers and soles had been given out in skins and sides, leaving it to the shoemaker to cut wisely and economically. What he could save by honest or dishonest means he felt free to keep in his "cabbage" stock box under his low shoemaker's bench. Old shoemakers are found now in every locality who recall with bitterness certain individuals whose dishonesty and poor work led to the development of the central shop. They blame the open way in which scrap leather buyers went about on regular routes to buy up the accumulated "cabbage" stock which meant loss to the capitalist while it was a source of profit to the shoemaker and presumably to the scrapman. These old people make the complaint with little realization that the boot and shoe industry has simply followed the example of the silk and the woolen industry in experiencing, in the course of growth, the need of saving stock as well as time, and of giving to agents the power of inspection formerly exercised by

---

[1] Johnson in his Sketches of Lynn, pp. 23–24, describes the ten-footers in detail. They came into use about the middle of the eighteenth century. Their size varied from 10 × 10 feet to 14 × 14 feet. The average size, however, was nearer 12 × 12 and 6½ clear in height. The garret, called a cockloft, was left unfinished and was the common receptacle of litter and everything not wanted for present use or wanted only occasionally. This garret was reached by a perpendicular ladder and was seldom cleared out. The number of benches varied from four to eight, according to the size of the ten-footer, each bench being placed far enough from the wall or the next bench to allow the arms to be swung out as the shoemaker sewed with full length thread. Such a space was called a berth, and the group of men who worked there was called the shop's crew. These last names are interesting as local color of a seaport, for in other shoe centres away from the water I have never found either crew or berth used in connection with a ten-footer.

Rise of the Central Shop: Its Function

The Central Shop System developed rapidly after 1820. The stock was cut there and portioned out ready to deliver to workers to do the "fitting," *i.e.*, the work on the uppers, siding up the seams, binding, and counter and strap stitching. When this process was completed, the uppers were returned to the central shop and given out with the proper number of roughly cut soles, as well as a definite quantity of thread, to "makers" who would last and then sole the boots and shoes. The makers had to wait generally for their work to be inspected or crowned at the central shop before new work was given out. The volume of trade and the amount of specialization can be understood by the study of a single month's business dealings in one firm's ledger, for 1833.[1]

Changes in Organization and Phraseology

The transition to this second phase from the first under the Domestic System was of course as gradual and almost imperceptible to the shoemakers and shoe merchants themselves as it was to the onlookers then, and as it is to research students of today. The account books changed their phraseology but slowly to keep pace with actual changes in business organization. Take, for simple illustration the word "making" as applied to boots and shoes. Roughly dated, until 1815 it meant making the whole boot. After that, when processes were subdivided and parcelled

---

[1] White and Whitcomb of East Randolph (now Holbrook) were employing in 1833, from September 4th to October 4th, twelve men and women to do fitting, thirty men to make, and one man to do all the cutting, pasting, and blacking. Both the styles and the finish were crude, and the usual proportion of labor was one central shop worker to about forty domestic workers. The larger sales during that month were to the following firms:

		No. of pairs
J. C. Addington and Co.	$226.18	250
Porter and Tileston	205.64	212
B. C. Harris	57.50	50
C. W. Howes and Co.	257.50	250
C. W. Howes and Co.	237.40	262
N. Houghton and Co.		218
Wm. Tiffany	100.00	100

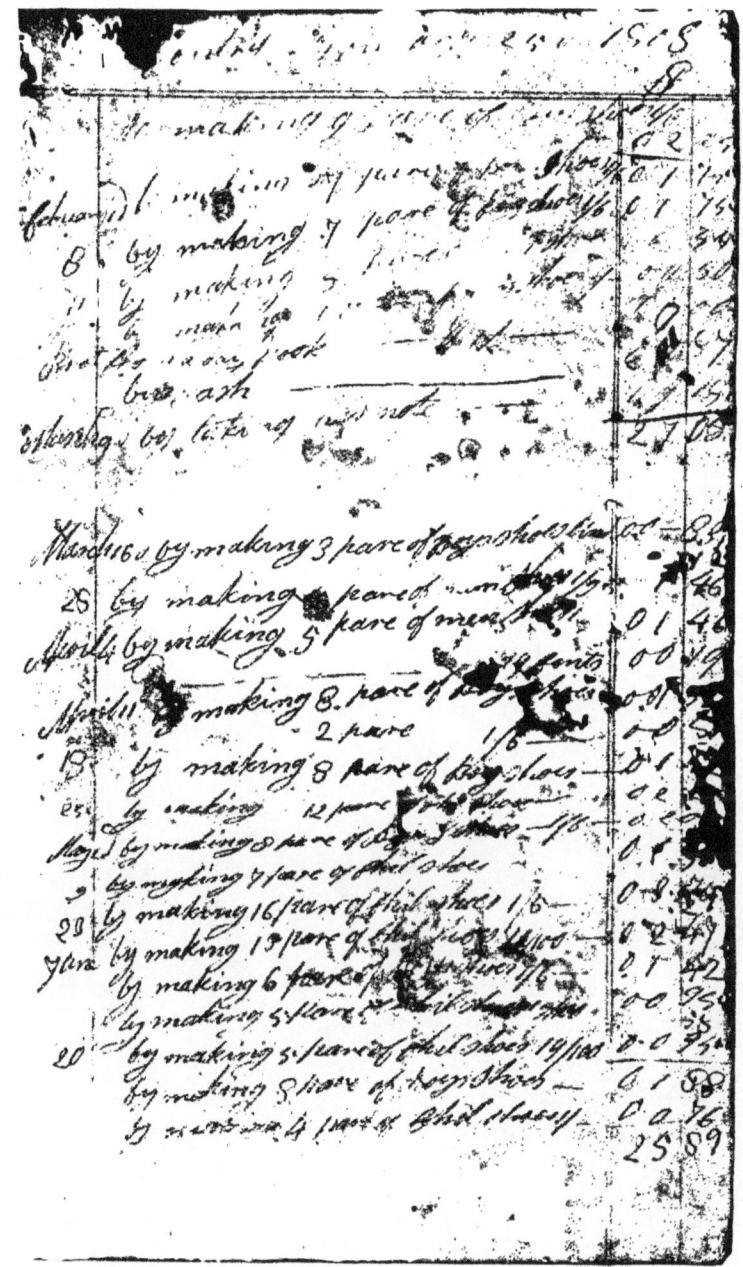

EBENEZER BELCHER'S ACCOUNT BOOK SHOWS PRICES OF SHOEMAKING

out to different individuals, the word came to have a technical meaning, and was meant to include only the lasting and bottoming. Where it appears frequently in one and the same account book, in one and the same handwriting, it is often hard to tell which is meant. The best guide so far has been to note the absence or the presence of other words designating supplementary processes, like siding up, and crimping, to decide when making meant only putting on the soles.

To show this transition not only in wording but also in industrial organization, certain accounts of individuals who manufactured boots and shoes both before and after 1810 have been reserved for consideration here, as the major part of their work was done in the second phase of this Domestic Stage.

Account Books Show Prices and Volume of Output

In an old ledger of a shoemaker-farmer of South Randolph,[1] beginning with 1807, the entries are continuous for about 40 years. They tell of the work [2] done in a "ten-footer" by Ebenezer Belcher with his regular apprentices and journeymen. The pages of the same account book tell of his simultaneous working of his farm with the aid of his sons. The development of the shoemaking industry in Randolph already has been traced [3] through the extra sale phase of the Handicraft Period into the first phase of the Domestic System, when the entrepreneurs appeared to organize the shoe industry among the Randolph shoemakers in order to meet the probable demands of the export trade. Silas Alden, Moses and Thomas French had been pioneers in it. By 1810, there seem to have been several merchant employers in the vicinity who were allies or competitors of Silas Alden. Three entrepreneurs, Isaac Thayer of West Randolph; John Arnold [4] of

---

[1] A settlement now known officially as Brookville, Mass.

[2] This account book was given to Harvard University in 1908 by his daughter-in-law, and can be studied there. A picture of this same "ten-footer" and of Ebenezer Belcher's house is attached to the fly leaf.

[3] Cf. pp. 18, 19.

[4] John Arnold was a farmer who ran a cider mill and sold shoes. He had no central shop or store, but he took the shoes to Boston in saddle bags as Quincy Reed did. Later he employed Amasa Clark, as one of his agents, to take out stock to domestic workers, to collect, and then to take it to Boston markets or to jobbers.

Braintree, Randolph's parent town; and Daniel Faxon [1] of South Randolph together with Bass and Turner who had a general store in West Randolph, seem to have ordered or taken most of the shoes made in Ebenezer Belcher's shop until 1820. Silas Alden had his central shop on the main street of the central village of Randolph. The ledger records show:

Silas Alden Dr. to Ebenezer Belcher

1807
Aug. 21 To making 15 pr. Boy's shoes.................. $6.25
Aug. 31   "   "   18 " men's   " 7.50
          "   "   7 "   "   b. t. shoes........... 2.92
Oct. 24   "   "   12 "   "   "   " 5.00

In this bill for making shoes, stock was not included. Silas Alden paid Ebenezer Belcher with meal, by cash, and by two orders, one on Jonathan Wales for $11.00 and one on Drake for $13.00.

By 1817 Bass and Turner, who had the chief general store in the town, were employing Ebenezer Belcher steadily to make shoes. They must have been furnishing the stock.

1817
Mar. 7 To making 20 pare of men shoes 2/............ $6.67 [2]
Mar. 20   "   "   50 "   "   "   "   " 16.67
Apr. 1   "   "   68 "   "   "   "   " 22.67
Apr. 8   "   "   5 "   "   "   "   " 1.67
          "   "   20 "   "   "   "   " 6.67
May 27   "   "   10 "   "   "   "   " 3.34
June 10   "   "   20 "   "   "   " 1/6 .... 5.00
June 24   "   "   20 "   "   "   "   " 5.00
July 11   "   "   40 "   "   "   "   " 14.00
July 15   "   "   15 "   "   "   "   " 3.75
July 19   "   "   25 "   "   "   "   " 8.33
Aug. 16   "   "   20 "   "   "   "   2/ 6.67
Aug. 16   "   "   " "   "   "   " 1/6 .... 5.00
Sept. 11 "   "   40 "   "   "   "   2/ 13.33

This is the account of two "runs," to use our modern phrase, the summer and the winter seasons' supply. The average price for making a pair seems about 33⅓ cents but the average output is

---

[1] Daniel Faxon had a store and gave out leather to be made up. He also took shoes to Boston in saddle bags and on his return brought flour and other groceries.

[2] Shilling here equalled 16⅔ cents.

DOMESTIC STAGE — PUTTING-OUT SYSTEM 47

harder to estimate accurately. The period of fourteen days between March 7 and 20 shows an output of 50 pair from Ebenezer Belcher's workers, but between June 10 and 24 only 20 pair are the apparent product. The number of apprentices whom he had that spring the ledger does not settle. Some years he kept two apprentices and from four to six journeymen. There are the names of eight different men being paid by Ebenezer Belcher for making boots between 1807 and 1809. Much of the work ordered from him in 1817 by Bass and Turner was probably done by other neighboring shoemakers in their own domestic shops, or in the ells or kitchens of their houses. Meanwhile Ebenezer Belcher was taking orders from other shoe merchants and for custom work,[1] besides doing mending. The accounts would seem to show that while the merchants furnished the stock for the big orders, Belcher was providing it for custom work and perhaps for some of the smaller orders. He bought his leather, sides of calf skin generally, of John White of Braintree and paid him in cash from $10 to $15 at a time.

```
 1807
 Oct. 19   Account of John White and Eben. Belcher by 5
              sides of lather............................ $13.50
              to 16 lbs. of lather........................   1.60
              "  2  "    " oil...........................    .50
 Nov. 12   "  5 sides of Neets lather 17/................  14.16
      30   "  4  "    "    "     "    17/................  11.33
           "  2  "    "    "     "    17/3...............   5.75
 Dec. 19 by 6  "    "     "     ".......................  17.50
```

Evidently Ebenezer Belcher was doing considerable custom work then and possibly sale work. If he had been furnishing the leather for the shoes at Bass and Turner's in 1817, it might be thought that he was doing belated sale work and leaving it on commission, or having it credited outright to his account. The ledger pages show definitely, however, that he was paid only for work, and generally only the one process [2] of making. There is no

---

[1] Seth Turner Dr. to me (Eben. Belcher) Sept. 3d, 1808.
    to 1 pare of women shoes 7/   1.75   paid
    to 1 pare of women shoes 7/   1.75
[These prices include stock and labor. B. E. H.]

[2] An exception to this came in 1818 when he charged Bass and Turner for cut-

48   ORGANIZATION OF BOOT AND SHOE INDUSTRY

evidence that he was ever an entrepreneur or used his ten-footer as a central shop. He remained a master shoemaker doing work under the Domestic System. As early as 1810 when Belcher was working for Isaac Thayer, of Randolph, the latter was furnishing the stock for fairly large quantities. From May 20 to August 11, Belcher made 158 (22 + 14 + 22 + 100) pair of men's shoes for him.

He sometimes sold both farm produce and stock to his journeymen,[1] and to his domestic workers. Samuel Ludden, who lived only a few hundred rods away from Belcher's house and worked on shoes pretty regularly for him, was sometimes credited with doing "15 pare" at a time. He had an account in this ledger where he was debtor to Belcher for vamps and soles, flax, knives, and leather taps and for Belcher's horse "to Lincorns." [2]

Entrepreneurs in the Shoe Trade and Their Central Shops

This same Lincoln's corner store, at the cross roads (from Bridgewater to Boston through Braintree, and from Randolph to

ting for them 100 pairs of boots. In later years, from 1836–40, his ledger pages show him doing work on uppers "fitting them," as well as "making." As this was generally women's work by that time, he may have been employing some members of his family or neighbors to do it.

[1] Ambrose Hollis to me, 1807

To 1 lb. of flax, to 1 half Bushel Corn .... $	.74
" meat & Butter................................	.36
	1.10
By making 11 pare of children shoes...............	1.20
To 1 peck of corn..................................	00.27
By making 6 pare of children's shoes...............	1.20
	.93
By making 8 pare of children's shoes...............	1.60
To 1 half bush. of corn to 1 lb. of flax.............	.74
	.76

[2] This was to Lincoln's general store in the centre of the East Randolph village. By 1821, Ludden was working on his own account for the Littlefield shoe merchants in East Stoughton (Avon) and often hired Eben. Belcher's horse to return work and get more stock.

Weymouth), was a beginning of the large Holbrook shoe manufacture and trade which flourished in the last seventy-five years of the nineteenth century. At first as a general store it dealt in sale shoes. Then, as early as 1816, Ephraim Lincoln bought up or had made for him quantities of shoes and took them into Boston in a two-wheeled hand cart made for him for this purpose. He was one of the first entrepreneurs who was not a shoemaker. He got the stock in Boston, brought it back for "return cargo," and had it cut up in "his shop" in the grocery store and "put it out" to be made into shoes. He developed the central shop idea and finally the manufacturing end of his business was of more importance than his general store. His experience was fairly common and typical in many towns of Massachusetts. His even more interesting and successful contemporary, Quincy Reed of Weymouth, who has been quoted as a typical early shoe entrepreneur,[1] marketed and produced a far greater supply of shoes. Reed's brother, Harvey, seems to have proved the organizer of their trade, and Quincy, of their manufacture. The firm continued until 1833.[2] In the years between 1809 and the year of the firm's dissolution, the business was large relatively and positively.

Other Interests and Ventures of Shoe Entrepreneurs

The duties of the Reed brothers in their Boston store and Weymouth shop in the early days of business (1810-15), included cutting out stock in spare moments; cobbling to repair customers' old shoes; making new shoes by the pair for individuals, by the dozen pair for jobbers; and selling supplies like blacking and leather, awls and tacks, to shoemakers. This fact their simple bookkeeping shows on scores of pages. In the decade from 1820-30, their correspondence, bills and receipts, and ledgers as well, give an idea of other more complicated duties and dealings. When Harvey Reed bought a whole Maine township, Quincy

[1] See pp. 13, 14.
[2] This date is particularly interesting to students of the development of the trade, for it is the very year when the Batchellers of Brookfield took up manufacture for that same Southern planter trade along the seaboard and the Gulf, and up the Mississippi River.

Reed withstood the land lure and had for a side issue a grain business, which kept four schooners [1] busy in the coastwise trade, and with Spain. The files of bills of lading show the details of this enterprise.

<small>Sales were made in the cities of the United States between Boston and New Orleans. Branch houses were established in Richmond and New Orleans, where business up the Mississippi River was immense. They also sold largely of certain kinds of goods to the West Indies; and in return made large importations of sugar and molasses. Mr. Harvey Reed was a man of remarkable ability. He had the care of the outside business of the firm, attending to its larger interests in the South and elsewhere. I think he was one of the original promoters of the Union Bank of Weymouth and Braintree, and of the Weymouth Savings Bank. Having become largely interested in Maine lands, he removed to Bangor in 1833 where he carried on for twenty-five years a large lumbering business until his death in 1859.[2]</small>

These facts about the banking and land interests have been added here, even if they exceed the limits of our period to 1837, because Mr. Harvey Reed's interests as an entrepreneur shoe merchant are so typical. Not only the Littlefields of East Stoughton, but also the Aldens of Randolph, the Batchellers of Brookfield, the Newhalls in Lynn, all in turn becoming leaders in their communities in bringing in banking facilities and in investing in distant lands, whether in Maine or Texas or Louisiana, sometimes taking them over for debt but oftener as an investment for the large profits gained in their shoe manufacture and trade. The Littlefield brothers of East Stoughton (now Avon), who were making money in the Cuban and New Orleans trade, invested over $80,000 in Aroostook County land before 1836. They "pulled out before that money was lost" in the general hard times of 1837, because they needed ready money for the New Orleans trade. There the business amounted to $100,000 a year between 1828 and 1837, but in the early 40's, the results of mismanagement and the general effects of the financial depression combined

---

[1] These were the days of simple transportation facilities. In 1812, Noah Thayer, of Randolph, drove an ox team to Richmond, Va., with shoes made up for such an export venture. He brought back a load of corn and cotton.

[2] This is quoted from J. W. Porter, who interviewed Quincy Reed in 1885, when he was ninety-one years old, and published the facts in the Bangor Historical Magazine, vol. I, August, 1885.

to bring bad debts, so that the firm had to be content to take payments in lands in Mississippi and Texas just when land was a drug on the market.[1]

The Littlefield Brothers as Typical Entrepreneurs in the Shoe Industry

These Littlefield brothers of East Stoughton began their work as entrepreneurs in the shoe manufacture and trade a little before 1815. There were four of them, Nathaniel, James, Isaac, and Darius.[2] They divided the duties of capitalists, manufacturers, and salesmen between them. Mr. Darius Littlefield used to drive into Boston Monday and stay until Wednesday, and again on Thursday and stay until Saturday, "at Wilde's Hotel" as it seemed necessary to add. Mr. James Littlefield was a travelling salesman. He never tired of telling of the hard times of 1816–17 the "froze to death" year, when the crops were so short that he took pay in the South for his shoes in flour and sold it for a good price and big profit up North. He had Philadelphia and Virginia trade, and sold brogans to the far South, even to New Orleans, and over in Cuba. They transported them in hair trunks about the size of a present day wooden case or box for boots, instead of the casks used for earlier trade. Casks were sent still, however, to Cuba, whether this was because they were better adapted to ocean freight or to get the hogsheads there ready for return cargo of molasses, it is hard to determine. They had, like many other entrepreneurs in the shoe business then, a grocery store. Mr. Darius Littlefield was taken into the firm in 1821. By 1822 he had got an appointment as postmaster, for, as he naively confessed in later years, his New Orleans mail was heavy and as postmaster he could frank his own mail.

By having the grocery store where his goods, bought at wholesale, could be paid out at retail prices to his domestic workers,

---

[1] Cf. p. 52 for more details.
[2] All this information was given to me orally in 1908, by Mrs. Varnum, who was the grandchild of Darius Littlefield. She was getting it from her father Cyrus Littlefield, then living at the age of eighty-six years but too feeble to talk to a stranger. From his reminiscences and answers to direct questions she was able to gather much material of value for this study.

or on orders from other manufacturers, he got rid of paying wages in cash. Mr. Nathaniel Littlefield moved to Randolph and built a house and later a shop, where he continued manufacturing as well as keeping up his share of financial interest in the East Stoughton shop where his brother Darius was in charge. When Mr. Isaac Littlefield died in New Orleans of the yellow fever in 1836, none of the family could go so far in the hard times to settle the business there, and it was left in charge of clerks or agents. Through them the firm tried to collect the New Orleans store debts and had to take lands in payment in many instances. In 1845 the firm of Littlefield Brothers failed in East Stoughton on account of their loss of the earlier well organized market for brogans for slaves, as well as on account of their uncollected bills. In 1846, Mr. Nathaniel Littlefield went South to see about properties there. These were in Mississippi and Texas and proved in the end to be a "dead loss."

The Littlefields and their Domestic Workers

During the years when the Littlefield firm was flourishing, from 1815 to at least 1837, the domestic workers came from miles around to the central shop to "take out" work; women got boot legs to side up and cord; men got boots to last and bottom for their own work, and straps to stick, tops and counters to sew on for their children's work. All the members of a family, oftentimes of a whole community would be found working on boots. The competition among the numerous workers, for now they did not need to be skilled shoemakers but oftentimes just men or women or children each knowing how to accomplish one process, made it unnecessary for the entrepreneurs to send out the work. Rows of beach wagons [1] stood along the fence before the Littlefield shoe shop many times a week, even as early as 1830. People had come from Abington, Mansfield and Middleborough, as well as from neighboring towns, like the Randolphs and Bridgewaters. Many an entry in the account books of Ebenezer Belcher of

---

[1] Beach wagons were two-seated, but could be made four-seated and the back leather curtain could be raised when the back seat was slipped under the front seat, so as to put in a case of boots.

South Randolph gives evidence of what neighboring shoemakers were "taking out" shoes, and later boots, "to make for Littlefield's."

Shoe Entrepreneurs in Natick, Mass.

A picture of the domestic shoemakers of Natick, Massachusetts, going for stock in this Domestic Stage, or having it taken to them, is given by Mr. Isaac Felch.[1]

The shoes were put out to be bottomed in little shops all over the community. The quarters were to be bound, the uppers closed, and the stock for it was carted to these workmen. My Uncle Asa was the first man [2] in Natick to make shoes to sell for the Boston Market. It was in 1836 or 1837. I have heard father tell of the Boston merchants coming out to meet Uncle Asa at the toll gates [3] of the Worcester Turnpike, to secure his shoes. The first time they failed to meet him, he thought the bottom had fallen out of the business.

This Isaac Felch's father gave up a contractor's transportation business between Lake Champlain and Boston, through Natick, with a six-horse team when the new railroad began seriously to take away his trade,[4] and joined his brother Asa in the shoe industry.[5]

Oliver Bacon, in his History of Natick, speaks of the construction of this Boston and Worcester Railroad in 1835, as giving a

[1] Mr. Isaac Felch wrote for me on his eightieth birthday, in 1913, an account of his connection with the early Natick boot industry.

[2] Both Bacon's History of Natick, p. 152, and Hurd's Middlesex History (vol. I, p. 554) say that the Walcotts did a shoe business for outside merchants, *i. e.*, "sales shoes," before 1835, beginning in 1828. This is disputed by the Felch brothers. Bacon (p. 153) says that Isaac Felch began business in 1836 and for the next five years employed seventy-seven persons in making and preparing the shoes to be made. During this time the average number of shoes made was 31,200 pairs. Bacon quotes figures for the volume of the John B. Walcott business in 1835 as 4050 pairs; 1836 as 11,000 pairs; in 1837, as 8310 pairs.

[3] Seemingly where the present Back Bay of Boston is located.

[4] By a curious incident, he put the axe at his own tree — for he contracted to bring the iron rails for the Albany Railroad, that were used for the tracks through Natick.

[5] Asa Felch made shoes first in an ell, cutting out, making up about sixty pairs at a time, and taking them to Boston to sell. Then when Isaac Felch, Sr., joined his brother, they made brogans in a barn on their land, in a room upstairs. Mr. Henry Felch has his father's big cutting knife now. It was only the cutting and crowning which was done in that barn-loft-factory, for the work was put out, as described above, to domestic workers.

new impulse to the shoe business, because it cost less thereafter to transport both leather stock and finished shoes and a new era began for Natick. This town had been relatively late and slow in the developing of its industrial conditions, as compared with other shoe towns. In 1805 "it had a population of 700, and scarce a common road worthy of the name, and hostility between the north and south ends of the town was at its height." By 1825, comforts began to increase; desks for the school house appeared, and the roads were gravelled. Most of the people were industrious and frugal farmers. By 1835, just when the shoe industry of Natick went out of the Handicraft Stage [1] into the Domestic Stage, there was "energy, action, new life in every direction. . . . The stage coach and ponderous ox-wagon had passed. . . . The light chaise and spirited horse were seen rolling along over the newly laid out turnpike." Of course this new life and growth are the characteristics common to all Massachusetts towns at this era of the prosperous inflation years which preceded the panic year of 1837. Yet Natick survived those years of financial depression with greater ease, perhaps because her manufacturing ventures were just beginning and there was not enough spare capital to be put into the larger outside investments which engulfed financial undertakings [2] of other towns. Natick's career as a boot town, however, except for the beginnings made by the Felch and the Walcott firms, belongs to the last phase of the Domestic System from 1837 to 1855, the days of Henry Wilson's boot-making career.

[1] Just what particular person or event was the impulse here is not recorded. I find a tradition among men over seventy-five years of age that a certain Frenchman whose name "began with Le——" came from Beverly and taught Henry Wilson how to make brogans. Was he the link between the new Lynn-Beverly Domestic System and the old regime where Uncle Eph Whitney was the "Cobbler of Natick"?

[2] Barber in his Historical Collections of Massachusetts gives the shoe business of Natick in 1837 as 250,650 pairs of shoes, valued at $213,052.50. Males employed, 263; and females, 189. Cf. Appendix XVIII.

DOMESTIC STAGE — PUTTING-OUT SYSTEM

The Shoe Trade of Brookfield, Massachusetts, and the Batchellers as Entrepreneurs

Brookfield was still another town which had its rise and rapid progress as a boot and shoe centre in this period from 1810 to 1837. The early years of the town's life while it was called Quabaug, during its Home Stage and Handicraft Stage in the shoe industry, have been described. By 1810, it was still in the town economy stage of finance and industry, with no export relations. That year marked the advent of Oliver Ward and Tyler Batcheller, shoemakers from Grafton. From contemporary account books of a general provision store of that village, we get a picture of the town life into which they had come to project their shoemaking activities.

General Economic Conditions in Brookfield

The week of June 19 to 26 of 1813 seems to have been either a gala time in the Brookfield villages, or else the opening week of Skinner and Ward's general provision store.[1] Some accounts in the firm's ledger opened during those days with a charge for "calicoa," or sugar or salt, but most of them with a charge of "1 pt. Brandy" or "2 pts. N. Rum." There were two hundred and fifty-eight pages in one ledger bearing the first date 1813, each with a new account. These pages represent that same number of customers and give an impression of sales made to people from outside the Brookfield villages. The next eighty-eight pages take entries for 1814 and 1815, while the earlier pages carried on the first customers. The index[2] shows the names of seven hundred and sixty different customers, and supposedly the heads of so many families. Rum, candles, tea, sugar, spices, gin, and salt

---

[1] The Skinner and Ward books were owned in 1908 by Mr. Henry Twitchell of Brookfield and treasured under the eaves of his house.

[2]

50 A's	2 E's	25 J's	10 N's	47 S's
88 B's	27 F's	10 K's	16 O's	14 T's
53 C's	14 G's	15 L's	22 P's	18 U's
48 D's	104 H's	43 M's	74 R's	80 W's
			Total	760 names

are the most familiar and recurrent entries in every one of these personal accounts. Sometimes a looking glass, or twine, or paper, some silk, calico, and bedticking are mentioned. The payment [1] is often in kind, in service, with only occasional credits of cash.

While many people in the villages of Brookfield were weaving for Skinner and Ward, they were probably doing custom work or bespoke work for their fellow townsmen. There was also a firm, known as the Branch Cotton Manufacturing Company, which was doing a good business with the Skinner and Ward store. Messrs. Dolliver and Norwell were selling shoes (August 9, 1813) to the amount of $62.26 and (October 9, 1813) $14.54 and (October 27, 1813) $30.60, to Skinner, Ward and Co. Here were sale shoes for a local market sold to stray customers, not to jobbers.

From another book showing the stocktaking returns of this same Skinner firm in July of 1814 and June of 1815 I can draw, for myself at least, a pretty clear picture of that store. The entries show the range of prices, the variety of stock,[2] the amount of stock on hand. The notes [3] receivable were mostly on short time (three months). The size of the accounts receivable, pure credits not secured by notes, running up to $5256.38 for that year, was seldom more than $50 for any one individual. The firm of Darius Hovey and Co. had run up a bill to over $100 and John

---

[1] Mr. William Upham paid his store bill with corn and by weaving shirting. In November of each year, a hog, weighing from 168 to 190 pounds, was credited to his account. The remainder of his indebtedness was cancelled by cash. Here we see the only evidences left to us of one William Upham, who in the industrial life of the community figured as a Handicraft master or Domestic worker. He was a farmer with the by-employment of weaving.

Calvin Davis, meanwhile was weaving and fulling cloth. Sometimes (April 9, 1814) he was weaving 79 yards No. 81 and 52 yards No. 82, which were credited to him for $9.69. Again (May 3), he was credited for $6.37 in return for 51 yards of bedticking which he wove for Skinner and Ward. This is the way he paid for snuff, sugar, garden seed, molasses, yarn, and milk pans, as well as for his shoes; also for a hat purchased during those same weeks. The fact that he paid in part by a hog, and cider, and bought garden seed and milk pans, suggests that he was a weaver-farmer, doing extra sales work on venture or else as a domestic worker for the orders of Skinner and Ward.

[2] Stock valued on that date at $1902.49.

[3] Notes receivable valued (with interest) at $385.76. Cash on hand $220.20.

Ward, a manufacturer, owed $135 on July 11, 1814. The debts payable by Skinner and Ward at that date amounted to $3360. Therefore, stock and cash on hand and notes receivable and accounts owed to Skinner and Ward Co. more than balanced the amount of the firm's indebtedness. That some of the stock was practically worthless or at least unmarketable is suggested by finding "seven old Fur hats and three felt boys' hats" in one annual inventory after another.

Rise of Organized Shoe Industry in Brookfield

Up to this decade (1810-1820) the community had been dependent upon its agriculture mainly, although some weaving was being done for other than home consumption. The shoemakers were attending only to the demands of definite customers, with occasional sale or "extra" shoes for the local stores. When in 1810 Oliver Ward had come to town, and by his work and enterprise, coupled with that of the Batcheller brothers, made Brookfield a typical shoe town, its shoe industry passed out of the extra sale shoe phase of the Handicraft Period through the period of specialization for certain markets [1] and of the rise of the central shop where stock cutting and inspection, as well as packing and shipping, could be done under the supervision of the entrepreneur.

The first shoes Batcheller made were chiefly of a low-priced quality, especially adapted to the southern trade. He packed these in empty flour barrels, consigned them to a middleman who ran a line of sailing packets between Boston and Havana, and realized a good profit. He continually, though cautiously, enlarged his business, both his central shop and stock, number of workers and customers. In 1824 he took his brother in as partner.

The Batcheller Brogan for Export Trade in South and West

The firm, before the 30's, had added to their business the manufacture of "Batcheller's Retail Brogan," a shoe adapted to the New England trade and "kept for sale in all the stores in this and many neighboring towns." Their main business was still the manufacture of brogans for the southern and western States.

[1] See Appendix XVI.

58   ORGANIZATION OF BOOT AND SHOE INDUSTRY

Meanwhile both the factory and the firm had been enlarged. By 1831 the new firm introduced the manufacture of the *russet brogan* for the southern slaves, the first that were made in Massachusetts, and this became their leading product.

Summary of the Development of the Brookfields as a Shoe Trade Centre

Here then, we have another simple though fairly graphic account of the rise of entrepreneurs of the Domestic System in the first stage. The Batcheller brothers entered the lists in the last phase of the Handicraft Period in Brookfield, with the idea of sale shoes not as mere extras in local stores, but as a venture for export trade to the South. With capital accumulated from the first ventures, they built up a business which outgrew a shoemaker's shop with its small group of apprentices and journeymen, and embarked upon the Central Shop, specializing for a definite market. Before 1835, they were putting out stock to workmen in their small shops in Brookfield and in most of the towns in the vicinity, in some instances the stock being carried to a distance of twenty or thirty miles. Besides this, they had a jobbing house in Boston, and regular customers in the South and West.[1]

The volume of business done by this T. and E. Batcheller firm in 1832 was 65,000 pairs, to the cash value of $52,000. The firm increased its output steadily. Meanwhile another North Brookfield firm, Deming and Edwards, organized a large business in 1835. In 1837, they moved into a three-story shop which they built in the centre of the village, and specialized in russet brogans and coarse thick boots designed for the southern trade. That same year they established a boot and shoe house in Mobile, Alabama, where they had already a considerable trade with the planters along the Tombigbee River. The slaves for whom these planters were buying shoes, had "feet of enormous size," and their measures, marked Tom, Pete, Sam, Joe, etc., on slips of paper, were sent North with the orders, and then returned tucked inside the respective shoes. Even in the hard panic years, 1837–39, this new firm was not only expanding, but even flourishing,

[1] See Appendix XVII, for details taken from Batcheller accounts.

probably because of two important facts — (1) it was not loaded with old, uncollectible debts; and (2), it could get the customers of the firms who were failing and going out of business.

The Development of Southern, West Indian and South American Markets for Shoes

Perhaps no one autobiography has furnished more light on the southern and West Indian shoe trade than that of Seth Bryant.[1] He gives with vividness of detail and closest familiarity the facts concerning not only the distant markets but the manufacturers and jobbers who supplied them. He naively claims that all Lynn manufacturers made the store of Mitchell and Bryant, on Broad Street, Boston, their headquarters from 1824. "Mitchell and Bryant had the *first wholesale boot and shoe* house in Boston. All the other dealers at that time were *jobbers* and kept all kinds of shoes. . . . Mitchell and Bryant specialized in *men's heavy goods*. Russet and black brogans were manufactured in Holliston, Mass., and we sold nearly all the shoes which were manufactured there, and also kip brogans and copper nailed shoes for the West India trade, made in Joppa." [2]

Lynn manufacturers were helped by the firm Mitchell and Bryant who sold only men's shoes to secure customers for women's shoes in which Lynn had long specialized. "We had a large southern and western trade with the grocery men, dry goods and hardware men. They could keep our goods and sell them by the case, and some of them could work off a great many cases in a year. . . . We helped set up young shoe firms out West.

"Messrs. Micajah Pratt and Nathan Breed of Lynn came to my store in Boston one day and said there were a couple of young men who wanted to open a shoe store in St. Louis. They were very capable, enterprising young men of undoubted integrity — without sufficient capital. They (Mr. Pratt and Mr. Breed) were going to let the young men, Hood and Abbott, have $500 worth

---

[1] Shoe and Leather Trade of the last hundred years by Seth Bryant. Published in Ashmont, Mass., 1891, p. 26.

[2] Joppa was a part of Bridgewater where Seth Bryant began life, and manufactured for many years.

each of their Lynn goods and wanted me to put in $500 worth of my men's goods. They said if Hood and Abbott succeeded they would be good future customers, so I (Seth Bryant of Mitchell and Bryant) put in $500 goods and got Joseph Hunt of Abington to put in $500 of his fine Abington goods. Hood and Abbott started then in St. Louis and became very successful merchants. I believe I sold Hood and Abbott a million dollars worth of goods after that. I presume Lynn manufacturers have started a great many others in the same way." [1]

Of all the jobbers and middlemen in southern markets, John W. Houghton seems to have had the most interesting career. He was born in Harvard, Mass., in 1787, and came to Lynn at the age of nineteen, fresh from his Harvard home to seek his fortune. He trudged to Lynn, and went to live with his brother-in-law with whom he served his time. Then he commenced manufacturing under the name of Jayne and Houghton in 1813. Business was not successful and Mr. Houghton, becoming somewhat disheartened but not discouraged, left Lynn. He went to Newark, N. J., and attempted to manufacture shoes but was not successful, so he took about $600 worth of goods and journeyed South to Savannah. When he arrived he had only twenty-five cents left, but he was hopeful. Taking his effects up the river to Augusta, he opened a store and eventually was successful.

He added gradually to his shoe trade so his establishment became a regular out-fitting store where planters from the country could obtain whatever supplies they desired for master and slave, "from a bell-top hat to a pair of peaked-toed shoes." The store in Augusta was in a crowded business quarter near the market house on Broad Street and supplied the trade with well-to-do planters of both South Carolina and North Carolina. Houghton's goods were being manufactured in the North, the most extensive orders coming from New England. Houghton bought a plantation twelve miles out of Augusta and erected "Yankee House"; laid out grounds in northern style and built a school house on the premises, containing 2400 acres, for his fifty negro slaves. By his will he gave freedom to these slaves and left $4000 to the executor

---

[1] Bryant, Shoe, etc., pp. 25–27. Not an exact quotation.

to use in defraying their expenses to the seaboard where the American Colonization Society would take them in charge to send them back to Liberia. He gave another $4000 for the erection of a school house for poor children of the parish to be known as Houghton Institute. Bryant in closing this naive account says that Houghton was one of the largest and best customers of Mitchell and Bryant's wholesale shoe business in Boston, buying russet and black brogans for the southern plantation, often lots worth $20,000 to $30,000 at a time, to be shipped to Savannah and up the river to Augusta.[1]

West Indian Trade

It was not alone for markets in our southern states that Bryant and Mitchell were manufacturing boots and shoes. They were supplying West Indian markets through Spofford and Tileston, jobbers of New York. These shoes were packed in Havana sugar boxes, 75 to 110 pairs to a box, and shipped from Joppa, Mass., to New York by way of Providence, R. I.[2] "I generally went on the same vessel which carried the shoes.[3] No steam boats were run till about 1824 and all water transportation had to be in sailing vessels. I sold for $1.25 per pair those shoes to Spofford and Tileston, who shipped most of them to Cuba." They bought also of Abner Curtis of Rockland, Mass., who used to drive into Broad Street, Boston, with four gray horses, his wagon loaded heavily with calf shoes and nailed brogans, all of which were shipped in schooners.

Samuel Train was still another New England entrepreneur who made a success of the southern shoe trade from his store at No. 28 Mercantile Row, Boston. "He started on foot from his home in New Hampshire for Boston, with his pack on his back, sat down under some elm trees in Medford and ate the last food in his pack and when he arrived in Boston, he was the possessor

---

[1] Cf. Bryant, Shoe, etc., pp. 9–18.

[2] As it was before the days of railways, the shoes were hauled overland from the manufactory in Joppa (near Bridgewater, Mass.) to Providence, without ever being shipped to the Boston store of Mitchell and Bryant.

[3] Bryant, p. 49.

of only fifty cents. He commenced business at the foot of the ladder. He did a large business, shipping his shoes in flour barrels, packed as nicely as crackers. He shipped to Charleston and Savannah. He became a large real estate and shipowner and was a man of excellence in every respect. He bought the place in Medford where he sat as a boy under the elms and ate the last loaf. He gave the trade much dignity." [1]

Mr. Bryant tells also of A. and H. Reed of South Weymouth and of Mr. James Littlefield of E. Stoughton who were manufacturing for the Cuban trade. The latter was so well known to the Cuba markets that Spaniards came to Littlefield's Boston office and made their purchase in person.

When Bryant and Mitchell somewhat later began to manufacture directly for the West Indian Trade, including Cuba, San Domingo and Hayti, they made a light calf nailed brogan. They sent 20 or 30 cases of shoes. For these Cuban sugar, brown and white, was taken in exchange. The white was shipped in turn to St. Petersburg, Russia, where Russian calfskins in an untanned condition were secured. Some of the brown sugar was shipped to Trieste and opium brought back for payment.[2] All of the shoes they sent to Hayti were traded for coffee which they imported in quite large quantities. These are evidences of the existence of more three-cornered trade which included New England and the West Indies.

South American Trade

With South America, the shoe trade had been developed by Augustus Hemenway of Boston, who had five or six ships running to Valparaiso, Chile, and bringing back cargoes of copper ore in bulk. "We would put up the shoes in nice little cases, twenty-four pairs in a case.[3] The cases were bound with iron straps. Hemenway said he could put two cases on to a mule's back and send them three or four hundred miles over the mountains. He kept his trade so private that he did not allow us to tell any one he bought shoes of us. He would send them aboard the vessel

---

[1] Bryant, p. 39.   [2] *Ibid.*, pp. 28–29.
[3] *Ibid.*: Shoe and Leather Trade of the last hundred years, pp. 28, 114.

after dark for fear other people would find out he was shipping shoes and he was just as private with other goods. He kept a gang of men loading the ship all night. The last bill of goods I sold him amounted to about $20,000 which I had his check for."

Altogether Bryant's picture of the markets and jobbers of the southern wholesale trade makes the research worker understand better the market conditions and expectations under which the shoe manufacturers of Massachusetts were producing and getting their successful markets in the twenties and early thirties before the Panic of 1837. All the export trade was full, evidently, of allurement and hopes. Some brought too great risk and no profit. Bryant tells of the successful ventures. One can imagine many unchronicled failures from overreaching in distant markets and unwise investments of capital, while others, like the E. and T. Batcheller, and the Mitchell and Bryant firms, held to their course with success.

The Second Phase of the Domestic Period Closes in Panic of 1837

Prosperity in and through the boot and shoe trade in Massachusetts towns was general in 1837. Not only individual shoe merchants were in the speculating, expanding mood; whole shoe towns were also. Lynn built forty-two new streets between 1831 and 1840. Just as Weymouth had needed a bank to help its shoe business, so Lynn had felt and met the need. The Lynn Mechanics Bank was incorporated in 1814, a savings bank in 1826, and the Nahant Bank in 1832, each with a capital of $150,000. Thus the shoe entrepreneurs, oftentimes the leaders in local progress and policies, had other problems than those of credit and markets, investments and tariff protection.[1] They had to help in town expansion.

Unfortunately not only town building on a larger scale, but the wonderful expansion of the shoe industry also, was to re-

---

[1] In 1816 the duty on boots was advanced to $1.50 a pair. In 1824 this amount was put on laced boots or "bootees" and a tax of 25 cents a pair put on prunella or other shoes and slippers. No reduction was made in these heavy protective rates until 1842.

ceive a check amounting oftentimes to complete disaster in the country-wide financial crisis of 1837. The large fortunes and the market as well came to a sudden end in the panic years of 1837–38. The money tied up in land investments could not be released in time to tide over the epidemic of failures. Then the Massachusetts Bankruptcy Law, by relieving debtors, settled the nicely poised fate of their braver, more far-sighted creditors. Many more manufacturing shoemakers had to fail. As a rule, the men who had smaller trade and smaller risks "held on" grimly, to become the backbone of a new phase of the Domestic System.

CHAPTER IV

DOMESTIC STAGE. PUTTING–OUT SYSTEM, 1760–1855

Phase 3, 1837–1855

Regaining of capital and of markets.

Growth of distinct boot and shoe centres.

Entrepreneur in boot and shoe manufacture makes that industry his sole business and not a side venture.

Capital turned over more rapidly and aided by opening of banks for discounts and loans.

Henry Wilson as a type of entrepreneur using small but active capital.

Eddy and Leach as a type of firm composed of a shoemaker allied with a capitalist, because competition of manufacturers demanded expert inspection of product.

New styles and a greater variety of shoes made in order to secure new markets and new classes of trade. Lynn leads on women's shoes — specialization.

Inventions and devices in Lynn and elsewhere to secure standardization of product and to economize time.

Rise of markets for boots. The Howard and French manufactory for boots, typical (1) of new variety of processes; of increase of styles and sizes, and in use of an agent for collecting and placing work among domestic workers; (2) of increasing degree of work done in Central Shop under supervision; and (3) of allied industries in order to use by-products, and to economize labor.

Summary of the Third Phase of the Domestic Stage in the Boot and Shoe Industry.

Domestic Period — Phase 3

Revival of Shoe Industry after Panic of 1837

For the relatively few boot and shoe manufacturers who survived financially the panic of 1837, there was a strong probability of holding their old markets at their own terms, because so few competitors were left. For many of the shoe merchants who had to fail, there was hope that their old knowledge, skill, and reputation might help them to regain and hold their old customers if they could get capital enough in this era of general depression and financial hesitancy to start their business again.

For new shoe manufacturers, tempted into the field by the reports of its "big profits" prior to 1837, in the hope of gaining the markets left open by the failure of other manufacturers, there was a good chance. People had to wear boots and shoes and there was wealth in the country, even if undue expansion of credit and investment had brought temporary depression and stagnation. Since the saying was as true then as ever that a merchant who has a market can secure capital, the vital thing for all would-be boot and shoe manufacturers, whether new or old in the business, was to secure markets, and by 1840 a new trade with new markets was gradually emerging from the old shoe towns. There was a growth of distinct boot and shoe centres, and a new type of entrepreneurs.

Growth of Distinct Boot and Shoe Centres

After 1837, the investing of capital in boot and shoemaking does not seem to have been a mere chance stroke of a sporadic entrepreneur. Even before 1837, it was a policy or custom for the business men of only certain localities to invest a relatively large amount of their capital in boot and shoe manufacture, and to employ a large proportion of the inhabitants, men, women, and children, in the manufacture of shoes. The table given in Appendix XVIII, compiled from facts and figures gathered in 1837-38 and given in Barber's [1] Historical Collections of Massachusetts, shows these proportions, and also the distribution of the shoe manufacture throughout the counties and towns of Massachusetts.

Localities in Massachusetts Adapted and Devoted to the Boot and Shoe Industry

Analysis by Counties. We have figures and facts also for the output of all other industries in Massachusetts in 1837. A study of them throws light on the question why, in certain localities, the boot and shoe manufacture, instead of the woolen and iron manufacture, attracted capital. The fertile soil of some counties,[2] made farming, with wool growing, the most productive industry. In the

---

[1] Barber seems to have been an indefatigable research student, gathering statistics at first hand in an age when people were beginning to demand and appreciate them, though the Government was not collecting and publishing them. He went from town to town, it is said, and personally gathered his figures and facts.

[2] *E. g.*, Berkshire, Hampshire, Hampden, and Franklin.

case of other counties,[1] it is obvious why the population was employed in fishing, shipping, shipbuilding, salt making, marketing sperm oil, or running iron furnaces. This was determined by their geographical position and their mineral deposits. Hat making had a firm hold and made a large output in several counties and tool making seems as general.[2] Worcester County which might easily have devoted itself as far as natural conditions were concerned to farming and grazing, and Essex, which was on its water line a fishing, shipbuilding and shipping community, were devoting the larger share of their capital and labor to boot and shoe making. The other counties devoted to the manufacture of boots and shoes were Norfolk and Plymouth. They were not so well fitted for productive farming or grazing, and never went so far into the woolen and cotton industry as the western counties and the Middlesex towns did. Absolute nearness or relative proximity through transportation and banking facilities to Boston must have made a very great difference in determining the choice of industry.

A study of the grouping [3] of towns as well as the natural resources of the counties and the respective towns leads to the conclusion that custom and the easy course of imitation were also factors in determining where capital as well as labor should be directed in Massachusetts. In short, the boot and shoe centres, according to the analysis of such statistics as are available, were generally in the vicinity of large tanneries and in regions near enough to Boston to enjoy banking and transportation facilities

[1] *E. g.*, Bristol, Barnstable, Dukes, Essex — having long coast lines.

[2] Local demand shows itself in the large manufacture of scythe snaths in the Western counties, shovels in the Middle, and shoe tools in the Eastern counties. In Brockton (Plymouth County) the manufacture of boots and shoes was a very important, though not the most prominent business of the town. The old iron industry was leading naturally into tool making for the farmer and the shoemaker by 1835. In 1836, Chandler Sprague was making lasts and boot trees, having purchased the right of using Thomas Blanchard's machine for turning irregular forms. Ten years later a long line of shoemakers' tools and devices were credited to the toolmakers of North Bridgewater. Even in 1835, North Bridgewater was more noted for the tools which it manufactured than for the product of the tools, *i. e.*, pegged brogans. It undoubtedly supplied the neighboring shoemakers for miles around with their tools.

[3] See Appendix XVIII for a summary of natural resources and products.

68 ORGANIZATION OF BOOT AND SHOE INDUSTRY

as well as Boston markets. If a man wished to invest in woolen and cotton mills, he went to Hampshire County or to Bristol; if in shovels and iron manufacture, he went to Bristol or to adjacent parts of Plymouth and Norfolk Counties; if in paper mills, he went to the middle counties of the state; if in straw hats, then to the middle counties or to Bristol and Worcester; if in shoes and boots, he went to Worcester, Essex, Plymouth or Norfolk Counties — already known far and wide as the boot and shoe centres.

Entrepreneurs Make Boot and Shoe Manufacture their Main Interest and not a Side Venture

Generally the capital for boot and shoe men after 1837, and the two or three years of strain, survival, and revival which followed the panic years of 1837–39, was no longer the "safely" reinvested savings of successful master-shoemakers and the hoarded savings of domestic workers who added thereto the savings from sales of their extra farm produce; it was oftener capital hired by entrepreneurs to secure the largest or the surest profits possible. If the venture was more likely to succeed in one town and county than another, the capital to be invested in the boot and shoe manufacture would be attracted there. The shoe business in 1837 to 1855 and the whole Factory Period which followed was not the outgrowth of village cobbler shops and "spare-time-spare-savings work for venture in export trade" as in 1810–37; it represented systematic and critical investment in seemingly lucrative business that would pay interest on the borrowed capital and yield considerable profits, not a mere livelihood or an extra source of profit to a merchant. The use of notes and the rapid turning over of capital suggested and made a demand for banking facilities.

Henry Wilson's Account Book for 1846

Henry Wilson of Natick was an entrepreneur of this new type. His account book[1] shows that he was turning over capital rapidly, aided by the use of banks for discounts and loans. His account book for the year 1846 is thoroughly satisfactory for giving an

[1] This account book is owned by Henry Wilson's nephew, Mr. Louis Coolidge, treasurer of the United Shoe Machinery Company of Boston, Mass.

DOMESTIC STAGE — PUTTING-OUT SYSTEM 69

adequate view of operations on a comparatively small capital in a Central Shop during this third phase of the Domestic Period. All the entries are made in his own handwriting and the pages include a full inventory of his personal property as well as his real estate for 1846. Prices of work, of stock, and of finished shoes are given. The geographical distribution of the product is deducible from the lists of customers with their addresses, and there is appended to each customer's name Mr. Wilson's opinion of his financial standing, "good," "fair," "failed." In the latter case a line was drawn through the name. The account of stock [1] taken that year reveals the custom of consigning boots and shoes, the manner of payment on the part of jobbers. The number and wages of workers whether men or women working in the Central Shop or at home, and the amount they did for Henry Wilson, can be compiled from this small account book. Figures from its pages seem to show this boot and shoe entrepreneur's financial condition at the opening in 1846 to have been good. He had property to the value of $5332.51 more than he was liable for. This included house, shop, stock of raw and manufactured materials, personal property and the notes due him. He was working with small capital, using many notes and turning over his stock quickly, yet he did a $17,000 business in 1838.[2] His output increased from 18,000 pair that year, his first year of manufacturing, to 58,000 pair in 1845; and from employing 18 hands in 1838 he raised the number to 52 in 1845. By 1847, he made 122,000 [3] pair, with 109 hands working for him.

[1] See Appendix XIX for the Wilson papers.

[2] To a manufacturer of today, used to dealing in far larger ventures, Wilson's venture must seem like a mere grain of sand, yet when one recalls that he had been born a poor boy in New Hampshire in 1812; was apprenticed for 11 years to a farmer in New Hampshire, and started his young manhood on a capital of two oxen and six sheep which the farmer gave him at the close of that term of years; that he had been a poor school teacher in 1837–38 and had learned the trade of shoemaking meanwhile as a sort of avocation, he seems not only successful and energetic, but typical of many Massachusetts shoe entrepreneurs.

[3] No explanation is given for this sudden and single large output. That of 1846 was 47,000 pair, while the year of 1848 had an output of only 63,000, and the number of hands decreased to 68. A similar peculiar great advance in output came in Wilson's town that very same year. John B. Wolcott manufactured 100,000 pair

The shoe which Henry Wilson manufactured was a cheap brogan for Southern plantation slaves, needing simply cutting, closing of short seams on uppers, and lasting, in order to make them ready to ship. His customers were in Boston, Hartford, Philadelphia, Baltimore, Camden and Charleston, S. C., Augusta and Savannah, Ga., Montgomery, Ala., St. Louis, and New Orleans.

A few sample pages (given in Appendix XIX) suggest the quantities in which these men bought and the prices they paid. They show also that brogans were being shipped by freight over the new Worcester Railroad, packed in boxes and no longer in casks or trunks as they were in the pioneer days of the southern and western trade. The account book pages show comparatively few names of individual workers. In 1846, six men appear to be at work bottoming or lasting brogans for Henry Wilson, while eight women were closing, *i. e.*, sewing up short seams of brogans. This work was done so often by the young boys and girls of the family and the pay collected by their fathers, that the wages paid to women and men represent the pay for the joint work of the family. Besides the six men who appeared as domestic workers, there were agents who, as "freighters," came to Wilson's central shop and took out work to be done. This they distributed in the countryside, and, collecting it in due time, returned it to Wilson and received the lump pay. Whether these freighters were paid a commission by the domestic workers, or a wage by Wilson for this collecting, is not apparent in the records. It was a custom in very general use in the boot and shoe regions and its inherent difficulties will be discussed later on. One entry shows a man bringing in 1830 pairs of brogans at one time. The number of domestic workers engaged upon this lot is not recorded. These brogans represented perhaps the winter's work of two or three men and a week's work of a large number.

of brogans in 1847, when in 1846 his output was 64,000, and in 1848 it had dropped to 84,000. For the next six years it averaged 100,000 again. Evidently new markets had been anticipated and secured in his case, while Wilson went out of business into political life.

Competition of Manufacturers demands Expert Supervision of Product

When Henry Wilson ventured into the manufacture of brogans, he was a shoemaker by training as well as a capitalist. He could be superintendent of his own shop as well as seller of his product. These were the years when capitalists who were not shoemakers, realizing the keen competition of such contemporaries as Henry Wilson, had the wisdom to take as partners men who knew the trade and could put brains against money. While in old days, the shoemaker-turned-capitalist could keep his workers in hand by very casual inspection or a wholesome fear of it, because he was known to have learned the trade, the "outsider," as capitalist, could hope at most to learn the stock, and did not try to master the processes. If he did not take a real shoemaker as a partner, he had to trust entirely to expert workmen, foremen and agents who were devoted to his interests, and let them determine the details of stock and processes, while he attended to markets and profits.

A typical case of this was the firm of Eddy and Leach, in Middleboro, in southern Massachusetts where there were scores of ten-footers. When "Deacon Eddy" who owned the general store and ran the postoffice of the town, decided to go into the shoe business and invest the $10,000 he had previously made in the shovel business, he took for his partner George Leach, who had only $200 to invest, but years of experience as a skilled shoemaker to add. Before this date, 1852, in a ten-footer on his farm, Leach had made up shoes taken out from various central shops in Middleboro and neighboring shoe towns, sometimes with the aid of just his own family, at other times employing women and men in their homes. The same 22×28 two-story building which had served Deacon Eddy for a store [1] and postoffice was thereafter known as the Eddy-Leach Manufactory. The upper rooms were used for storing stock and finished goods, for cutting upper and

---

[1] This store was in a central spot where the mailcoach on its way from Boston to Plymouth stopped each day. After the railroad linked up Plymouth and Braintree, and Fall River and Boston via Middleboro and Stoughton, the mailcoach was still the link between Plymouth and Middleboro. See map of Massachusetts facing page 14.

sole leather, and for inspecting and crowning the shoes brought back by domestic workers. Gradually here as elsewhere a gang to work at bottoming was employed. A 14-foot ell was added for a stitching room in the late 50's, where Mr. Leach's two sons, skilled shoemakers in the third generation,[1] did the stitching for the firm.

Specializing and Standardizing the Products

Skilled workmen who had no capital, nor executive experience which warranted their being taken into a firm, were much in demand if they were trustworthy and had initiative and adaptability as well as technical training. New and more numerous styles, and better workmanship than had previously been put into ready-made shoes, were the order of this new shoe period in the Lynn region. High legged boots, never before general in New England except in custom work, were the new departure of Randolph and the neighboring towns which became known as boot towns.

The Massachusetts towns which specialized in shoes, whether they were cheap coarse brogans or fine ladies' or cheap children's shoes, can be seen by figures presented in the table in Appendix XVIII.

Women's pegged and common sewed shoes[2] were manufactured chiefly at Lynn, Haverhill, Worcester, Milford, Natick, Randolph, Abington, North and South Reading, Danvers, Georgetown, Stoughton, Woburn, Weymouth, and Stoneham. Boots were made in each of these towns except Lynn, but were the leading article of manufacture in Haverhill, Milford, Worcester, North Brookfield, Spencer, Grafton, Randolph, Stoughton,[3] Weymouth, Abington, Hopkinton, and South Reading.[4]

[1] Cf. Chap. VI for more facts about the Leach shoemakers.

[2] New York and Philadelphia made boots and shoes of finer quality. Philadelphia led on account of large stocks of fine calf skins and morocco leather, and in number of its skilled German workmen. The annual sales of Philadelphia were about $15,000,000. Cf. U. S. Census of 1860, p. lxxii.

[3] Tucker Brothers of East Stoughton generally made up boots and shoes only on definite orders. They had a Boston office with samples to show buyers from the South and West, but had no retail stores as the Littlefields did, until Mr. Nathan Tucker went to Cincinnati in 1838 and sold boots and shoes made by various manufacturers in Lynn and Randolph. At that time Westerners did not take as fine a grade of boot or shoe as the Southerners. This was not so true after the gold mines opened up in California. [4] Cf. U. S. Census of 1860, p. lxxii.

In short, mostly women's shoes were made in Essex County, mostly men's boots in Norfolk and Worcester and Plymouth Counties.

The number of *boots* made in Massachusetts

in 1845 was 3,768,160
in 1855 was 11,892,329

The number of *shoes* made in Massachusetts

in 1845 was 17,128,411
in 1855 was 33,174,499

Brockton (North Bridgewater) had been making both boots and shoes in 1837, with emphasis decidedly on the boots. By 1855, the town had become decidedly a shoemaking centre.

	1837	1845	1855	1865
Boots	79,000 pr.	44,711	66,956	103,066
Shoes	22,300 pr.	115,476	694,760	1,009,700

Meanwhile, Randolph, though showing a clear tendency to go from specializing in shoes to boots, still made more shoes than boots.

	1837[1]	1855
Boots	200,175	345,100
Shoes	470,620	363,300

Lynn was apparently the first to reorganize on the new basis of standardization of product and better quality. It realized that the market was already making hints if not demands as to styles, instead of accepting quietly anything the shoemaker provided. This was a very real change from earlier conditions.

Need of Standardization

Johnson, in his Sketches of Lynn,[2] told of the helter-skelter work done on shoes in the booming days before 1837: —

There were probably more poor shoes made at this period than were ever made in our city before or since, in the same length of time. The stock for the most part was better than the workmanship, and the soles were generally better than the uppers. The great defect was a lack of system, which ig-

---

[1] Barber, Hist. Coll. of Mass., p. 482.
[2] Sketches of Lynn or The Changes of Fifty Years, by David N. Johnson, pp. 151–152.

nored all the laws of adaptation. Firm stout soles were joined to uppers that were evidently got up with no reference to wear; and worse than this, if both were equally good — as was sometimes the case — they were often spoiled in making up. As an illustration of this, thousands of pairs of boots cut from stout grained leather upper stock, and having soles of the best quality, were spoiled by the miserable expedient of using paper stiffenings. When it is understood that this paper was not the stout, compact leather-board of the present day, but a tender straw-paper that a drop of water would penetrate through and through, no comment is needed to demonstrate the utter worthlessness of the article when it was ready for the foot of the wearer. Nor was this the only unscientific and wasteful arrangement. Shoes were sewed in such a manner that they dropped to pieces long before they were half worn out; and when the sewing was good, the labor was wasted by the senseless practice of trimming the uppers close to the stitch, a practice [1] that made it impossible to wear a pair of these shoes a second time — thus causing a waste that could be reckoned by tens of thousands of dollars, if not by millions.

Causes of Poor Work in the 30's and of Improvement in the 40's

This lack of care was not all due, probably, to the feverish booming spirit which preceded the business panic of 1837. Other causes could be enumerated, such as (a) entrepreneurs who undertook manufacturing without being shoemakers and were unable to properly oversee and inspect; (b) shoeworkers who had never been apprenticed or taught the trade; [2] (c) the spirit of irresponsibility, felt by new manufacturers, had crept into the market, where if one customer would not buy, others would, and where the buyers [3] were competing instead of the sellers.

Competition of Manufacturers brings Improvement in Stock and Processes

After the hard business strain, the shoe business gradually regained its old markets and found new ones through close competition among the shoe manufacturing merchants, who had to tempt and coax trade with the same kind of schemes and wiles

---

[1] Would a shoemaker dispute the possibility and probability of this?

[2] *I.e.*, the marginal producers hired to help out in the competition for workmen in rush seasons and years.

[3] Post-war conditions of the markets with lessened production and increasing demands are demonstrating similar possibilities today which may serve to make people of 1920 understand conditions of 1835.

that one uses with a convalescent. New ways of cutting, finishing and packing shoes appeared. Laced shoes as well as buttoned shoes, and shoes laced on the side instead of front, became the fashion. A realization came gradually that the custom shoemaker need not be allowed to keep the monopoly of making so many shapes and sizes of shoes that all kinds of feet might be well fitted. It seems now as if Lynn shoemakers realized fully for the first time in the 40's that all four factors in shoemaking needed constantly to be taken into account. Not alone materials and labor, but patterns and tools counted much in adapting the stock of shoes to a wide range of customers. Up to this time, Lynn had either followed the styles current in England in the seventeenth and eighteenth centuries, or those in the United States in the nineteenth century. When its Latin-American trade started, Lynn suited the real or supposed fancies of South American and West Indian women for bright colors and fancy trimmings. This was its extent of adaptation to environment and to the human foot. Cloth for uppers, grained leather or kid for foxing [1] was the rule for women's shoes. A rough grading as to sizes and widths was known to exist though few pairs of three's and few "fulls" [2] were really mates.

Tin Sole Patterns for Sole Cutting

Johnson, the Lynn shoemaker-writer already quoted, describes in a heartfelt way, naive but savoring of real memories, the need of inventions to secure uniformity. Tin patterns for shaping soles made uniformity in the shape of shanks, though they abridged the independence of cutters. Around the shanks there had always been a "chance for the display of original genius. A single box of shoes, the product of three or four workmen, would display as many different styles of shanks. . . . The question of matching shanks, when the day's work was to be tied up, imposed a degree of responsibility not experienced at the present time. It was found that the range taken in a single day between the two

---

[1] The foxing is the extra or ornamental surface of leather over the upper of a shoe.

[2] Fulls were shoes cut to be full over the instep.

extremes of wide and narrow shanks was considerable, sometimes showing violent contrasts."[1] One wonders if the change to uniformity here was not primarily a move in the interest of time-saving for the manufacturer rather than merely one to relieve in the future a retail shoe dealer and his "fussy" customer, who might want her two shoes to match.

Patterns for Uppers

In 1848, George Parrott applied a principle, already known in turning lasts, to a pattern-making machine for uppers. This secured [2] (a) exactness in the proportion between the several sizes; and (b) made gradual increase in the fullness over the instep and around the ankle. This exactness was felt to be necessary thereafter, especially in the smaller sizes for women and in children's shoes. The first patterns had been made of straw board or some even less enduring material, and each manufacturer had cut his own patterns. Johnson adds that these "had not been characterized by mechanical exactness."

Shoes began in the 50's to be "crooked shoes." Before that they had been "straights" *i. e.*, no rights or lefts, though the knowledge of making crooked shoes had been possessed and used by ancient Egyptians, Greeks, Romans, Mediaeval and even early modern shoemakers as archaeological and printed materials prove.

Of course variation had to be made in both uppers and soles in making crooked shoes rather than straights, and a new shoe manufacturer without cut-stock on hand was more likely to adopt this new style.

Stripper and Sole Cutter

Already two other machines had been invented to make sole cutting more uniform than the tin patterns could keep them. The stripper, with its heavy blade worked by foot power, cut the sole leather into strips across the width of the side of leather, or in various widths corresponding to the length of the sole required. This gave *exactness of length*. The sole cutter itself which

---

[1] Johnson: Sketches of Lynn, pp. 17–18. [2] *Ibid.*, p. 20.

gave *regularity* of shape and *uniformity of width*, was the invention of Richard Richards, a Lynn last maker who patented it in 1844.

Allied Industries. Cut Leather

Even then, Lynn [1] was vitally interested in developing *allied industries*. Daniel Estes as early as 1845 made a business of selling cut sole leather, not outer soles, but inner soles and stiffening. Then in answer to demands for all grades of cut soles, of good as well as cheap stock, several other firms went into the business in Lynn.

Shoe Boxes and Cartons

The Lynn shoe manufacturers were among the first to realize that shoes were more attractive to retail trade and reached it in better order when packed in separate pasteboard boxes or cartons. In early days when Lynn shoemakers had tramped to Boston with venture shoes in bags slung over their backs, or had been fortunate enough to ride on horseback and carry the shoes in saddle-bags, they found no competing manufacturers with better ways of handling goods for market. From all the shoemaking towns, brogans and shoes of all sorts had generally been sent in barrels and those sent to the West Indies competed only with others sent in casks from England. Shoe boxes had not been generally used until the 30's.[2] It was in 1836 that shoe-box making on a large scale was established in Lynn by James Buffum, following up the success made by Benjamin Mudge and Elija Downing, who began as early as 1825. It was at the end of this decade of the 30's that paper shoe boxes were put on the market by Abner Jones of Lynn. Very few more were made before 1850, when George Cushman set up in the business, which proved a good one,

[1] Lynn and Brockton today are as famous for the allied industries of shoe manufacture from last making machines to rands, stays, and patterns, as for their tremendous output of shoes. Even by 1880, it was said in Lynn, "The trade of our city in cut leather reaches all over the country, and it is estimated that from one-third to one-half of the entire product is sold to dealers in other places." Johnson: Sketches of Lynn, p. 353.

[2] Accounts in Appendix XVII tell of the Batchellers' paying for boxes, but probably wooden outer cases, not cartons, in 1830–34. Cf. Ch. III for frequent mention of the use of trunks and casks and sugar barrels for packing shoes for transportation.

continuing all through the 50's, for then the practice of "double packing" of fine shoes had become general. The nicer the shoe, the greater the care taken in making and getting it to market since the stock and workmanship once put in were too valuable to be marred by carelessness. When the treeing and finishing had been done in the central shop, the stock was thereafter kept under careful conditions even in transportation, and proper precautions against dampness that might produce mould were deemed an essential part of plans for shipment.

Linking up of Inventions to Develop the Shoe Industry

It is worth while to follow thus in detail this development of system and uniformity in the line of specialization and standardization of the shoe product in Lynn, because that product was (a) necessarily finer in stock and workmanship in women's and children's shoes and slippers; and (b) was known to wider markets earlier than the output of footwear in any other part of Massachusetts. It seems to have led always, either because of the high standard set by John Dagyr in 1750, or because of this very characteristic of specializing in footwear for smaller and more tender feet. Lynn's early and superior technical advance seems to have been due to two factors: (1) the nature of its product's demands; and (2) the competition among its own manufacturers who had a large amount of creative power [1] along mechanical lines. It is no sporadic happening that made more shoe machine inventions on Lynn's borders even back in the 40's and 50's than anywhere else, nor that the first sewing machine, called the dry thread, used on uppers in Massachusetts, was installed in 1852 by John Wooldredge in his Lynn factory. He had an expert come from Philadelphia to instruct the first operator,[2] and by 1854,

---

[1] Lynn was even then what Brockton became in the 70's, a shoe city. They both continue to be such for shoemaking with all its ramifications into allied industries, like pattern making and shoe machinery, are in the very air, literally and figuratively, since sign boards on high buildings greet the eye as people come into the city, and every inhabitant is able to talk shoe manufacture.

[2] By a curious coincidence, this Lynn operator who was destined to take away Hannah's job of binding shoes (see Lucy Larcom's poem) was called Hannah also, Hannah Harris.

Nichols Connor, of Lynn, invented a binder to carry the galloon binding, doing it more regularly and faster than the skilled hand of any man or woman.

Tariff on Shoes

Whether Lynn influence at Washington had more to do than that of other shoe centres in gaining protection in the tariff of 1842 for their product, is not a matter of record, as it was when Eben Breed helped to secure certain favoring clauses in the tariff of 1789. It is noticeable, however, that while the tariff on men's boots and bootees of leather fell to $1.25 per pair,[1] and on men's shoes or pumps to 30 cents per pair, that the tariff of 50 cents on women's boots and bootees of leather; of 40 cents on children's boots and double-soled pumps; and of 25 cents on women's shoes or slippers whether of leather, prunella or other material except silk, remained as it had been in the Tariff of 1824. One feels that Lynn shoe manufacturers must have helped on this legislation by making known their desires or needs for protection.

Organization of Shoe Business in 1848–55 shown in Typical Account Books

For the progress made in Lynn as a shoe centre by the close of the third phase (1837–1855) of the Domestic Period, we have interesting documentary evidence. By good fortune, the Historical Society of Lynn has in its keeping some of the account books of Christopher Robinson and Co. for 1845 and 1855. Robinson had built a factory of brick on South Commons Street in Lynn in 1848, and his business, though small, was typical for the time and place. We have his note book, showing details of organization in the shoe business in 1848.[2] One realizes from these pages how few machines were then considered necessary for shoemaking; how small an amount of capital was needed to equip a Central Shop as late as 1854;[3] yet with such equipment, Robinson and Co.

---

[1] This was 25 cents less than in the earlier tariffs of 1816 and 1824.

[2] See Appendix XX.

[3] See Appendix XX; account of stock taken by Robinson and Company in 1854.

were furnishing goods to customers widely scattered over the country.[1]

Natick Specializing in Brogans, in 1855

Though Natick was not yet, in 1855, the shoe town that it came to be in the last half of the nineteenth century, its shoe manufacture far outran, in value and in number of workers, that of any other industry in the town. Some further remarks of a local historian[2] in 1855 give an insight to the progress made in Natick shoe manufacture as far as the division of labor and introduction of machinery were concerned.

> The purchase of leather, selling of shoes, and preparation of them for market are now the work of the manufacturer. The cutting, lining, packing of the upper leather, belong to another class of hands; of the sole leather to another; pegging is done either by machinery[3] or boys: lasting and trimming[4] by journeymen; binding and stitching by girls or by machinery, while polishing the tops and soles furnishes employment for two other sets of hands.
> Making the boxes in which the shoes are packed is another branch of the shoe business which affords employment for many hands.

The only part of this labor on shoes done regularly outside the central shop or manufactory was the binding and stitching when it was done by girls. The lasting and edge trimming was done by men working in gangs either inside the shop or outside in ten-footers.

---

[1] A partial list of customers in 1855 is the following:

Wheelock and Daniels, N. Y.	Grannis and Stewart, Albany, N. Y.
D. R. Hubbard, Rochester, N. Y.	Chas. Ramsdell, Buffalo, N. Y.
Stephen Oliver, Lynn, Mass.	Cornelius Sweetzer and Co., Saco, Maine.
John J. Ashby, Salem, Mass.	Talbot and Cunningham, Providence, R. I.
Charles Coburn, Boston, Mass.	Ira Cheney, Durham, N. H.
A. M. Haines, Galena, Ill.	Potter and Phillips, Davenport, Iowa.
White and Page, Richmond, Va.	Hendrick and Markley, Madison, Ind.
Rufus Elmer, Springfield, Mass.	

[2] Bacon was a lawyer of Natick, writing this history of the town. Cf. History of Natick, p. 152.

[3] The pegging machine had been invented in 1818 and put quite generally in use for brogans and cheap boots and shoes. Mr. Stephen Belcher of South Randolph saw pegged boots for the first time in 1827. In 1832 James Hall of North Bridgewater invented a machine for pointing pegs.

[4] Trimming means cutting and smoothing of edges of soles and heels.

All this development of Natick's shoe manufacture had come within the single third phase of the Domestic Period, roughly dated from 1837 to 1855. When Natick entrepreneurs began the first manufacture of ready-made shoes (1) the use of a central shop, where storing, cutting, and inspecting was done; and (2) division of labor or specialization in processes, were already customary factors in shoe manufacture. A coarse shoe, either a plow shoe or a brogan, was Natick's one specialty all through this period, and all the figures show the price to average one dollar.[1]

Rise and Development of the Boot Manufacture for Special Markets — 1837-1855

Just as Lynn and its neighboring towns in Essex County specialized in the manufacture of women's shoes bringing them to a higher standard by using far greater uniformity of technique; and as Brookfield with its Worcester County neighbors, and Natick with its Middlesex County towns kept to the manufacture of brogans in the 40's and 50's; so Randolph and many of her Norfolk County neighbors specialized in boots after the opening of the 40's,[2] though previous to that, figures show the preponderance of shoes, mostly brogans. By 1855, Randolph was as well known as the producer of high class boots[3] for California, Australia, and Texas trade, as New Bedford was for whalers' pumps; Raynham for its sailors' pumps for Cuba, its Balmorals for white artisans, and its plow boots for Western farmers. This whole development of specialties seems to have come mainly from the keen competition of producers for markets after the hard times of 1837, which, as has been said, had to be coaxed and exactly suited. The first boots, *i. e.*, long legged boots with sewed heavy soles, had been made in Randolph about 1830. By 1837, one old shoemaker told me, "plenty were being made." The examination

---

[1] See Appendix XXI.

[2] Though this specialty was confirmed in its youth by the large demand from Australia and California, it had been born before that time and thus was ready for vigorous growth when that demand came with the advent in California of former Randolph bootmakers who became gold seekers.

[3] The word "boots" here is meant for high legged, heavy leather boots, in contrast to low shoes or the ordinary boots of today.

of the account books bears this statement out if we do not think of "plenty" as "the majority." Statistics show that 200,175 pairs of boots and 470,620 pairs of shoes were made in Randolph in 1837. Here was a substantial beginning for the development of the boot specializing after 1837 when specializing meant success.

The Different Class of Customers Wearing Ready-Made or "Store Shoes" Called for Specialization

The Randolph boots were never intended to take the place of the cheap brogans or plow shoes, but of the custom-made top boots that had never been entirely out of use or style.

Curiously enough, it was the expansion to the West and the new frontier [1] that made the market and demand for the new kinds of boots and the better processes which manufacturers produced after the opening of the 40's. From the custom bootmakers in any large city like Boston, New York, Philadelphia, Baltimore, Richmond, Charleston, Savannah, Mobile, or New Orleans, all the rich and well-to-do men of the states north and south had ordered their own boots as a matter of course. If they were Southerners, they bought cheaply made brogans by the hundreds for their slaves; if Northerners, they expected to see farmers and hired help in the fields and shops and their own growing boys all wearing rough brogans or plow shoes. For themselves, they said they had to have shoes or boots made to order to get anything to fit or to be comfortable. These same men on the frontier, living away from access to the cities on the Eastern seaboard, had to get along without having new custom-shoes made for them. The young cities of the West would hardly have been remunerative places for custom shoemakers; a drummer or agent [2]

---

[1] Chicago, in 1840, had less than 5000 inhabitants; St. Louis in 1840 had more than 16,000; Cincinnati grew from 24,831 in 1830 to 46,338 by 1840. Detroit grew from 2200 people in 1830 to over 9000 in 1840. Columbus had but 6000 in 1840. The Southern cities, however, were already large, for New Orleans had 46,000 inhabitants in 1830 and Richmond had 16,000. Baltimore, one of the chief flour markets in the country, had a population of 102,313, and Mobile had 12,672 in 1840.

[2] See page 58 about the Brookfield firm of Deming and Edwards, with their established agent at Mobile. Nathan Tucker of East Stoughton, Massachusetts

with a line of ready-made boots and shoes from one of the Eastern States was much more likely to meet demands. When these drummers came they saw chances for different grades of workmanship and better grades of stock. If their home firms could make a satisfactory boot, though not on the individual customer's last but with the same style of workmanship and stock as the new man in the West would have ordered in a custom shop in the East, their firms might create and hold a new market.

Howard and French as Typical Manufacturers of Boots in Randolph

The pages of account books for the firm of Howard and French, in Randolph, reveal the technical progress and the increase of variety in stock, sizes, and workmanship that came in the 40's. Before their Californian and Australian trade began, they developed a local Massachusetts trade which was in the hands of their agent, Amasa Clark, and the account [1] of stock sold through him gives not only the variety of stock, but the range of wholesale prices. He had a regular route down in the Lynn-Haverhill district, where men's boots and shoes were not commonly made. He sold these customers large bills of goods and his dealings were mostly with jobbers or general store keepers, for he practically always sold a dozen pairs of each kind at a time. Single pairs, or signs of retail sales were rare exceptions after 1845. Clark collected payment in the form of notes, and occasionally in cash, from the customers. Entries [2] in the books crediting him with turning over much money give us an idea of the relative importance of his work.

When Amasa Clark was at home in Randolph, he used his horse and wagon for taking out stock to domestic workers and

went into the boot and shoe manufacture in 1836 with his brothers, who were already in business there. The customers, mainly in the South, along the seaboard and in New Orleans, were taking a high-grade boot. In 1838 the Tuckers started a retail store in Cincinnati, where Nathan Tucker sold the shoes and boots of various manufacturers in Lynn, Randolph, and East Stoughton. Before that, 'the firm generally made up definite orders only, and in the Boston office they kept samples to show the buyers who came from the South and West.

[1] See Appendix XXII. [2] *Ibid.*

bringing it back completed. All Liberty Street [1] bootmakers depended on this means of transportation for their stock and finished product, and Howard and French were saved the delays by which other firms suffered and lost. Mr. Clark's own interest in getting stock ready for his trip down to Lynn probably made him just the right person to go in and out among the workers, for he thus acted as a sort of unofficial foreman in "getting the run" through. This was necessary, for the irregular number of days of work paid for by Howard and French suggest the truth of the tradition that men took time off for planting and haying, as well as for occasional fishing trips.

Typical ledger pages [2] with entries for the three months from January to March, 1845, give details of Howard and French's customers and sales, which furnish a good idea of the volume of this firm's business, such as the study of Amasa Clark's sales alone could not give. Other pages, giving records of stock bought by the firm, furnish other means of realizing the volume of their business. This was small compared with the Walkover or Douglas concerns of today, but large compared with that of even the late 30's in other boot and shoe towns.

Wages. Truck System Survivals

These accounts show not only the greater cost of the finer grades of stock used in the 40's for boot and shoe manufacture and the higher price paid for boots by wholesale customers, but the higher wages paid by manufacturers for labor. Instead of compiling the wages, however, in the gross for any specific three months, it would seem more effective to take them throughout ten years (1845-55) to illustrate certain points such as specialization, piece and weekly wages, the wages of men and of women, of skilled and ordinary workers, of labor on high-priced and on low-priced goods.

By way of introduction, there are traces of the survivals of the truck system of paying wages, which was common before 1837

---

[1] See map, p. 18. A settlement lying between South Braintree and the junction of Liberty Street with Main Street, Randolph, where the Howard and French factory was located.

[2] See Appendix XXII for these pages.

when banking facilities were scarce and inadequate. The day of payment by orders on stores,[1] or truck wages, was nearly over for Randolph shoe workers by 1845, for entries in ledgers show that payments were almost invariably in cash. Only occasional mention is made between 1845 and 1855 of such articles as butter, potatoes, apples, and hay as payment.[2] As there was already a well established bank in Randolph where notes could be discounted and checks cashed, the firm did not have the old excuse for delaying payment. The payments were still made irregularly, ranging from $1.75 at a time to $10.00, just as the domestic workers on return of boots, or the cutter in the factory, asked for some cash. This was, of course, long before the time of a weekly pay roll.

Piece Work in Homes and Ten-footers

Cutting, treeing, and varnishing were always done in the factory and represent the work which might be supervised directly by the firm, partly before the stock was sent out to domestic workers to fit and make, partly after it came back to be "finished" by treeing and varnishing, and to be inspected and crowned. There were meanwhile scores and scores of men and women regularly employed, directly and indirectly, by Howard and French to make and fit boots in their homes. Gideon Howard, who lived over in South Randolph, near Ebenezer Belcher's old shop, took out stock and brought back finished boots. He had a "gang" over in his twelve-footer who fitted, made and finished: one lasted, one pegged and tacked on soles, one made fore edges, one put on heels and "pared them up," and in case of hand-

[1] The usual form of orders on stores read: — "Please deliver to the bearer goods to the amount of——." Dated and signed by the manufacturer or his clerk.

[2] Mr. French of the firm of Howard and French had a Vermont farm, and probably payment in farm produce was a mutual advantage. When Howard and French began business in 1840, they had paid generally in orders on the store of French and Spear, evidently a good sized general store. The entries in the day book for just three months show the following wide range of articles taken on orders in lieu of cash wages: boot-trees, long sticks, sponges, starch, shoe boxes, nails, grindstone, planes, plank, canvas, silesia, silk, vest pattern, trimmings, awls by the gross, paper by the ream, fur hats and almanacs, cashmere, sheep skins, as well as a lot of second hand clothing valued at $15.00. As late as 1855, two of their workers, Nathan Freeman and Elbridge Jones, took part of their pay in groceries.

sewed shoes, two or three sewers were needed to keep the rest of the gang busy. The "gang" had become a distinct factor in shoemaking. As business grew larger in the 40's and the 50's, the demand was greater for ten-footers to replace kitchen ells, and for separate seats in them. Some enterprising men in Randolph invested their capital in making little shops to rent in toto or by the seat at so much a week. All the men who worked in such a shop "chipped in" to pay the cost of the fire. These groups of men in a ten-footer gradually took on a character due to specialization demanded by the markets with higher standards and need of speed in output. Instead of all the men working there being regularly trained shoemakers, perhaps only one would be, and he was a boss contractor (not to be confused with men in Philadelphia who were called garret bosses and employed from one to twelve workmen in their own rooms, buying their stock and selling their product to jobbers and retailers), who took out from a central shop so many cases to be done at a certain figure and date, and hired shoe workers who had "picked up" the knowledge of one process and set them to work under his supervision. One of the gang was a laster, another a pegger, one an edgemaker, one a polisher. Sometimes, as business grew, each of these operators would be duplicated. Such work did away with the old seven-year apprenticeship system. In this gang system lies the genesis of the factory, for it has the essential characteristics, specialized work under supervision under one roof. The regulation price for fitting, *i. e.*, putting on binding, siding, or sewing the seams of uppers, was seven cents a pair, nearly double the price paid for fitting brogans. Ideas of the wages for such domestic workers and for shop workers can be gathered from these same Howard and French books.[1] A difference in piece wages for larger sized boots can be traced after 1847, as well as for work on finer grades. There are full records of women's wages, as domestic workers.[2]

[1] See Appendix XXII.   [2] *Ibid.*

Working Days, Wages and Specialties of Workmen in Central Shops

(a) *Day Wages.* Although there is no weekly pay roll to consult, a fairly accurate estimate of weekly earnings of the men working in the central shop can be gotten from the ledger pages. Nathan Freeman worked by the day as cutter in the shop at the regular price of $1.00 a day.[1] His wages amounted to $293.50 in 1854. In April of 1855, his pay began to be $1–1/3 a day, and it was the same till 1857. This wage was paid mostly in cash, at intervals of two or three days. It was paid in part, however, in cheese, flour, butter and leather stock.

Elbridge Jones was also working by the day at $1.66–2/3 per day in 1854. His record[2] shows that he worked about $24\frac{1}{2}$ days on an average a month, 293 days a year. His wages in 1854 amounted to $468.34. This sum was also paid partly in cash, partly in dinners at the rate of 16–2/3 cents apiece, in stock, in butter and in "hay for Jones's horse."

(b) *Piece Work Wage in Central Shop.* In 1855 Sylvanus Pratt was treeing in Howard and French's shop. It was his specialty and he, like the other four[3] men employed there as treers, did 100 pairs a week on an average for seven cents a pair for regular length boots. Short-legged grained boots were easier to tree and the price was correspondingly smaller.

[1] Number of days worked by Nathan Freeman:

26 in Jan. 1854.	$20\frac{1}{2}$ in July, 1854.
24 " Feb.	25 " Aug.
27 " Mar.	26 " Sept.
22 " Apr.	$24\frac{1}{2}$ " Oct.
25 " May	26 " Nov.
24 " June	23 " Dec.

[2] Number of days worked by Elbridge Jones:

21 in Jan. 1854.	19 in July, 1854.
24 " Feb.	26 " Aug.
25 " Mar.	25 " Sept.
21 " Apr.	24 " Oct.
22 " May	22 " Nov.
26 " June	26 " Dec.

[3] Ira Howard, Edwin Howard, Luther Rowe and Henry Bangs. See Appendix XXII. A tree is a form over which a shoe or boot leg is cleansed and polished in the factory just before it is packed.

Luther Rowe treed and varnished and earned more than Pratt did in any one week. He was young and vigorous in the 50's, and one of his fellow-workmen has told me that Luther always put in a "long day" when he could. This, he explained, was their expression of working as many hours as there were of daylight in summer and winter. Evidently the weeks when over $7.00 was earned contained some of the "long days" of work, even in winter time.[1]

(c) *Cutters and Stitchers.* There was a rise of wages for some shop workers in the later 50's; Francis Wayland Alden was working by the day in the shop, cutting at a

    wage in 1854   of $1.00 per day
    "   " 1855–57 " 1.25 " "
    "   " 1857   " 1.33¼ " "

Jerome Fletcher became an expert stitcher on the wax thread machine later. He worked at a day-wage until 1859. In 1860 he began on piece work there in the same shop. Records show such items as

    257 seams at   4c.................................$10.28
    64 cases turned 10c....................................6.40
    151 seams   4c...6.04
    64 cases turned 10c....................................6.40

(d) *Crimping or Forming.* Thomas Mackedon was forming or crimping, *i. e.*, shaping the vamps or fronts of long-legged boots. This was a dirty, hard kind of work and received relatively high pay. "Once a crimper always a crimper" was true not only of one workman's life, but of his sons. The stain on his hands made

---

[1]

	1855	By treeing			LUTHER ROWE, Cr.			
Oct.	6	106 Pr. Boots	$7.42	Nov.	17	102 Pr. Boots		$7.14
	13	100 " "	7.00		24	23 doz. varnished		1.44
	20	85 " "	5.95			100 Pr. Treed		7.00
	20	85 " "	5.95	Dec.	1	81 " "		5.67
	27	108 " "	7.56			5 doz. varnished		.31
Nov.	3	96 " "	6.72		8	108 Pr. Treed		7.56
		12 " "	.84		15	114 " "		7.98
	10	93 " "	6.51		22	117 " "		8.19
					29	102 " "		7.14

Note uniform rates of 7 cents a pair for treeing, and of 6¼ cents a dozen pair for varnishing.

him appear a crimper every hour of the week, even on Sunday. Mackedon's shop, a ten-footer on Union Street, was the scene of crimping of tens of thousands of boot legs.[1]

(e) *Special Work on "Californians."* When the Californian trade began, Howard and French got their share and paid even higher prices for work done on boots for that special market.

```
1850
Apr. 11   By 12 pr. Californian made  .......................  $14.00 [2]
    15    "   "   "         "       "  .......................   14.00
    18    "   "   "         "       "  .......................   14.00
    24    "   6   "  Long Sewed     "  .......................    4.75
May 12    "  12   "         "       "  .......................    9.50
    12    "  12   "         "       "  .......................    9.50
```

Before that, the highest entry for other workmen on "12 pr. Californian" had been $10.00. Amos Kingman did some of the finest dress boots that Howard and French manufactured and Jonathan Wales, their agent in San Francisco, was glad to take them, for he sold them at retail for $10.00 a pair.

March 7, 1849 is the first entry on Howard and French's books of this so-called Californian, and one is surprised to find it so early, yet it shows the eagerness with which shoe manufacturers were adapting themselves to new needs or anticipating possible demands perhaps by cleverly naming a style of boot fit for the original transcontinental (a) counters journey of each 49'er.

Allied Industries and Economies of the Boot and Shoe Manufacture

Not only did labor have to become more specialized and efficient in this recuperating period of the 40's and in the early

---

[1]
```
        THOMAS MACKEDON WAS CREDITED ON
1855
Jan. 20  By Forming
              18,900 pr. boots to Jan. 4/55 .......................  $850.50
              i. e. .04½ a pr. price
1856
Jan. 14  By Forming
              21,446 Pr. boots to Jan/56 .......................  965.07
    For 1860, the bill runs up to $1066.
```

[2] Note that "made" here means simply lasting and bottoming at a price of over a dollar a pair.

50's, the third and final phase of the Domestic Stage, but it had to be economized to meet closer competition. And not only labor but stock also had to be economized by the manufacturers. These two motives are behind the buying of counters [1] and soles already cut. In old days, workmen fashioned counters and often innersoles from leather scraps. Howard and French, in 1855, were buying counters of N. M. Capen, in fairly large quantities, e. g., enough on September 25 to provide for nearly 1800 pairs of boots.

There was a Phinney Company, in the neighboring town of Stoughton, making counters as early as 1845.

> In 1845, Sylvannus C. Phinney commenced in Stoughton the manufacture of counters for boots and shoes. These are and always have been made of leather by this firm. Previous to this, counters had usually been cut out at the shoe manufactories and were fashioned by some one in the sole leather room in charge of that department. Mr. Phinney appears to have been among the first to realize that this great industry was in the future to be divided into many departments. So hard was it to convince manufacturers of Massachusetts that anything useful or profitable could come of specialties, that barely fifty sides of sole leather were cut per week, that being then considered a very respectable and satisfactory business. From the small beginning of 2500 sides of sole leather cut in 1845, in 1879 it had become necessary to cut up 90,000 sides.

(*f*) *Leather Shoe Strings.* It is more than likely, however, that the reluctance of the shoe manufacturers to buy counters came from the feeling that there was spare leather and labor in their own cutting rooms to be utilized, but just about the time that Howard and French began to buy counters, they began also to sell scrap leather. For this, J. Winsor Pratt of Randolph was a purchaser. He had set up in the business of leather shoe string making [2] and by supplying him, Howard and French transformed

---

[1] A counter is the stiff piece of leather around the back of a heel between the lining and the leather of vamp.

[2] One shoe string could be made from about a square inch of leather by sticking an awl or a knife point firmly through the leather and a board beneath it, and pulling the scrap slowly around to meet the edge of another knife. Thus a 27″ string was cut in a spiral out of the scrap, which was otherwise waste except for fuel.

what they had deemed a waste into a by-product. They occasionally bought strings of him as the accounts show.[1]

Howard and French were buying lasts of Moses Linfield, who was credited in the first three months of 1854 with 53 lasts of various sizes.[2] Thus there was an increasing amount of specializing among manufacturers concerned in boot making and allied trades in Randolph as in the Lynn region.

By the close of this period, North Bridgewater (Brockton) had, like Lynn, developed industries allied to the boot and shoe manufacture on a larger scale than any other shoe centre. There are figures from a local census in 1855 to show this:

Occupation	No. employed	Occupation	No. employed
Shoemakers	420	Shoe tool manufacturers	2
Bootmakers	134	Box manufacturers	1
Shoe cutters	37	Awl manufacturers	3
Shoe manufacturers	21	Patent leather makers	3
Boot manufacturers	10	Boot treers	6
Last makers[3]	7	Currier	1
Awl makers	13	Trimmers	5
Boot formers (crimpers)	6	Blacking makers	4
Boot tree makers	2		
Shoe tool makers	9	Total	685

[1] Account of Howard and French with J. Winsor Pratt.

```
       1854                               J. WINSOR PRATT
Apr. 17  Amt. forward ........... $26.37   1854
     19  136 Pcs. calf pieces ......   6.83   Nov. 10  by 100 bunch strings ........ $29.00
         125  "  grained pieces .....   4.05
May  15   67  "    "    "    .....   2.01   1855
          54  " calf   "    .....   2.70   Jan.  8  By cash ..................... 50.00
     19   1 boot box ..............    .37   Apr. 18  "   "  ..................... 20.00
     31   42 lbs. ground peas ........   1.26            Use of horse and wagon .....  2.75
          39  " calf pieces ..........   1.95
Jan. 22   96  "  "   "    ..........   4.80   Dec.  1  200 bunch strings ........... 58.00
       1855
Mar. 29  Amt. Brought up ........... 149.90
Apr. 25  160 lbs. Calf Pcs. ..........   8.00 (5c.)
         110  "  grd.  "  ..........   3.30 (3c.)
May  11   77  " calf   "  ..........   3.85 (5c.)
          50  " grd.   "  ..........   1.50 (3c.)
```

[2]
```
   1854 Jan. 17   5 P & L Lasts 3/9 ................................... $3.13
        Feb. 11  11 "  "    "    "    .......................................  6.88
        Mar.  7   6 "  "    "    "    .......................................  3.75
        Mar. 14  11 "  "    "    "    .......................................  6.88
             19   6 "  "    "    "    .......................................  3.75
             24   5 "  " J. H. 70 .........................................  3.50
             24   1 "  "           3/4 ....................................    .64
             25   8 "  "    .............................................  5.00
                 ──
                 53
```

[3] These seven men made 40,000 lasts valued at $10,000 that year.

These men with the help of domestic workers from within or outside of the town, made 66,056 pairs of boots and 694,760 pairs of shoes, valued at $724,847, in 1855.

Relation of Machines to Economizing of Time and Labor

The development of technical processes, with the necessary tools, devices and machinery for boot and shoemaking through all its stages of industrial organization, has been touched only incidentally in these chapters so far, even in this one when so many new devices and a few important machines were being added not only to the list of inventions, but also to the average manufacturer's stock in trade. Just enough mention has been made in this particular chapter, describing, as it does, the organization of the shoe industry in the third phase of the Domestic Stage, to explain the rapid drift of shoe manufacturers to standardization of product and to economy of time and labor. Yet the development of even such machines as the wax thread or the dry thread for stitching uppers rapidly and regularly to economize labor, or the strippers and block lasts for getting uniformity of product, though much more obvious, was not nearly so vital and effective as the almost unnoticed drift towards factory-like supervision of labor on boots and shoes in securing both standardization [1] and economies of labor and of stock.

[1] Old workmen did not see this point then or in later life. To them it was a trick of a particular manufacturer to make more money somehow out of his workers, or a tempting chance of making better money for themselves. The reason for the transition from a laissez-faire policy to central shop inspection of work was not very apparent to them.

Henry P. Crocker, of Raynham, 85 years old in 1912, described one case of this transition. The finishing of soles had been done outside with little uniformity by the individual "makers" who put on the soles. When the Gilmore Factory took the process into the Central Shop, they gave a man only one cent a pair for finishing. Several men had failed to make a living at it and Crocker hesitated at the offer, in 1847, for he was getting $9.00 a month "making" at home for some one else. Soon however, he was earning $1.00 a day, then $2.00 and sometimes $3.00. His "partner" working beside him, did the sand-papering, Crocker put on the oxalic acid to whiten the leather, polished it with a sheep's leg bone, stamped it with the firm name and gave it to another man to "dress."

Review of Use of Tools in Earlier Shoe Manufacture and the Use of Machines Before 1855

Up to about 1840, the shoemaker had used mainly just such tools as had been used for centuries. His kit, already described, included a lapstone and hammer for pounding leather, awls for pegging holes, a stirrup for holding the shoe in place on the knee, pincers to pull the leather over the last, nippers to pull out tacks, bristles for needles, and hemp thread for sewing, a buffing knife or scraper for the sole leather, and a shoulder-stick for polishing soles. Then there came a little skiving machine, run by hand, and not very satisfactory to the older men, accustomed to skiving with a regular knife. The next machines to be invented for boot and shoe work were the stripper, for cutting up sides of sole leather, and a leather-rolling machine, which came in 1845 to save both time and strength formerly used in hammering the sole leather on a lapstone.

Because the market was already making hints if not demands as to styles instead of accepting quietly anything the shoemaker provided, the use of different shapes and widths of block lasts came in the early 40's. Shoemakers no longer depended upon "instep leathers" for making "fulls" and "slims." Substantial patterns came into use. Sole patterns which gave uniformity of shape and width at ball and shank, and patterns for rounding the soles after they were stitched, were invented. Irons for polishing the edges came into use. Heels were put on women's shoes again,[1] and men began to specialize in heeling. Several styles of pegging machines,[2] and a machine for cutting up pegs had been patented and put in general use by the time [3] the sewing machine, invented by Howe in 1846, had been adapted to upper leather work on

---

[1] Spring heels, which took the place of high ones in 1830, had given place to no heels at all. Brogans were almost heelless, having but one lift. Cf. Johnson, Lynn Sketches, p. 340.

[2] Cf. Rehe, pp. 180, 189, for the story of Krantz and the introduction of the pegging machine into Germany.

[3] Proofs of dates of invention are in an unsatisfactory state at present. Not only local historians and biographers, but the United States Census Reports of 1860 and 1900, give merely approximate or relative dates for many inventions and for their introduction to practical use.

shoes by John Brooks Nichols in 1852.[1] There was the "dry thread" machine with a shuttle and two threads for the lighter upper work, and the "wax thread" to do chain stitching for the heavy work of "siding up" bootlegs. These sewing machines even then impressed people with their significance. Instead of merely making things easier or a "bit more speedy," they produced work which could not be matched by hand in either speed or appearance.

Summary of Third Phase of Domestic Stage

To give a brief summary of the development, characteristics and attendant results and policies of the third phase of the Domestic Stage seems worth while, even if it involves repetition.

In 1840, a new trade with new markets was gradually emerging from the old boot and shoe trade, but it had to be coaxed. New styles, niceties and novelties in processes required greater specialization. No manufacturer could afford to lose a single customer by slipshod work or poor stock. The competition for employment gave the employers a chance to choose only the best workmen. Only those shoemakers were in demand who had developed a reputation for a specialty,[2] like pegging, crimping, treeing or finishing.

More workers were taken into the central shops [3] and expected to do one thing "up to the standard," *e. g.*, they were hired to tree boots, or to finish bottoms. Not only every sole in a case must be uniformly finished, but the appearance of the shoes in all the cases sent to one customer must be the same. Competition led to

---

[1] For details concerning Nichols and his adaptation of the Howe Machine, see the Sewing Machine Journal of April 25, May 25, 1904, July 10, July 25 and February 10, 1911. These are the first-hand reports and illustrated. They give valuable facts on the Howe-Singer-Leavitt machine controversy.

[2] Samuel White, of Randolph (b. 1831), was known as a champion rapid pegger. Spectators came from neighboring towns in the late 40's to watch him work in his little ten-footer on Union Street. He has told me that his speed was gained by using both hands at once running the dink, using the awl, and hammering in pegs in quick succession, while he held his supply of pegs in his mouth. Cf. Case of Henry Crocker as finisher, footnote, p. 92 and Luther Rowe as treer, p. 88.

[3] Central shops were frequently spoken of as manufactories during the 40's and early 50's.

new styles of long legged boots to fit conditions of Western life, first of Australia and then of California in turn. Wholesale manufacturers attempted to capture the liking of men of Southern and Western frontiers, who had been used to custom-made shoes all their lives. It goes without saying that there were no custom bootmakers in Western frontier settlements to take individual orders. Riding horseback and walking over prairies needed peculiar styles adapted to their peculiar demands. So did digging in gold mines and climbing mountain trails. Old traditions were unsettled. The well-to-do man in Boston and Baltimore did not dream of wearing any but custom-made boots or shoes. The well-to-do man on the frontier in San Francisco or Melbourne, Australia, had come, however, by 1855, to supplying his needs at a retail shoe store with no qualm or scorn. The organization that had been built up for supplying the Southern trade with brogans was capable of enlargement; the whetted appetites of entrepreneurs were ready for more profits even with new risks. To get the shoes to distant points on time, to make them appear attractive enough to hold customers, and to keep all the work in a hundred cases of boots up to the standard of the sample, needed a new organization of manufacturing methods and processes. Specializing and labor-saving went hand in hand with standardizing. This increasing specialization led to the entrance into the shoe trade of young men and women who learned and knew just one process, and to the cessation of regular apprenticeship [1] for shoemakers.

Not only did labor have to become more specialized and efficient in this recuperating period of the 40's, the third and final phase of the Domestic Stage, but it had to be economized to meet closer competition. That fact, added to the demand for standardization of product, led to the introduction of more machinery into the boot and shoe industry. By 1855 so general had become the use of sewing machines that shoemakers, who could afford it, had them in their homes to use on both cloth and leather. But it

[1] There seems to be no definite, or even indefinite, case in Massachusetts of a shoemaker's becoming a regular apprentice to learn to make a whole boot after 1840. This has been a point in my investigation for several years.

was left generally for the manufacturer to put the machines into his central shop, or for the man with some capital and genius for machinery, who bought or leased the "wax thread" and "dry thread" machines, to set them up either in a stitching shop or in a central shop where space was hired. The more adaptable men and young women followed the machines into the shops, leaving the older people to "side up" and bind shoes by hand at home. Thus a new stage of organization came in the boot and shoe industry, bringing to an end not only the third phase but the main life of the Domestic Stage, where the "putting out" system had prevailed and the entrepreneur had worked in his central shop while the domestic workers labored in their "ten-footers." Only the "making," *i. e.*, lasting and bottoming, of sewed shoes continued to be done by domestic workers far into the next period, until the McKay machine for sewing soles and finally the Goodyear welting machine put an end to this last survival of the Domestic System.

CHAPTER V

FACTORY STAGE, 1855-1920

PHASE 1, 1855-1875

More direct control and supervision, with all workers in the factory, which is the old Central Shop enlarged or replaced by a larger building.
Chief Characteristic — Direct Supervision.
  (1) Origin of Factory Stage.
      Local explanations of the rise of the Factory Stage in boot and shoe manufacture.
      External signs of the new Factory System.
      Effect of Californian and Australian trade.
  (2) Development of Factories up to 1860.
      Transportation in the 50's and 60's.
      Financial crisis of 1857, and its effects on the boot and shoe trade; crisis feared but not experienced in Randolph. Lynn suffered some. Conditions in the Boston shoe trade. Boston shipping lists as indicators.
      Development of factory buildings for the boot and shoe industry. The Gilmore factory in Raynham. Lynn factories.
      New England boot and shoe manufacture in 1860.
      Massachusetts boot and shoe manufacture in 1860.
      Brookfield boot and shoe manufacture in 1860.
      Accounts of Kimball and Robinson, showing survival of Domestic System: probable reasons: later rapid development of factory organization: work book and order book under new firm name of H. E. Twitchell.
  (3) Development from Civil War to 1875.
      Effect of Civil War on boot and shoe manufacture. North Bridgewater's (Brockton) slow development into the McKay shoe specialty under factory conditions.
      Lynn's factory conditions at the close of 1865, and of 1875.
      E. and A. H. Batcheller's factory at North Brookfield as the largest in 1865 and 1875; accounts to show volume and details of their business; the contract system.
  (4) Summary of the rise of the Factory System and progress in boot and shoe industrial organization to 1875.

PHASE 1, 1855-1875

Rise of the Factory

SUPERVISION at the central shop of work done on boots and shoes, whether at home or in the shop, was the less obvious but the vitally important characteristic of the Factory Stage. Its need had been felt and met to a rapidly increasing degree in the close of the Domestic Stage. The Factory Stage did not come into existence in the boot and shoe industry because, as it is commonly supposed, the central shop was replaced by a larger building called a manufactory or factory, nor because of the installation in it of heavy expensive machinery, nor the use of power to run it, but because industrial organization, in order to secure uniformity of output, economy of time, labor, and stock, demanded foremen to superintend, and regular hours of steady work on the part of men and women employed in all of the processes of shoe-making.

Local Explanation of Rise of Factory Stage

While central supervision had been gradually discovered to be effective by individual entrepreneurs here and there, its presence as a recognized factor in the boot and shoe industry was not apparent to the shoemakers themselves until about 1855. Just which of the shoe centres had the greatest amount of factory-like organization first, it is hard to determine, though probability points to Lynn, where the industrial organization for the shoe industry had always matured both early and rapidly. The answer to the question depends upon the local experience of shoemakers, *e. g.*, in Randolph old shoemakers say that the Californian and Australian demand for boots, and Jonathan Wales's insistence on prompt delivery of ordered goods, necessitated changes in the system of production. Yet Brookfield, with its output of brogans for the South and Middle West, like Lynn, with orders for women's shoes, had in turn entered upon this Factory System gradually though definitely, without sharing in the impetus of Californian markets. It was well established throughout Massachusetts before the McKay machines, run by power, "took work away from men who had been domestic workers"; and before the Civil War, by "using men for the army, required their replacement by women who ceased to be domestic workers." Neither of these so-called explanations of the coming of the Factory System offered by several men when questioned are logical or true to facts. Though the causes of the introduction of the Factory

System in the boot and shoe industry were the same from one end of Massachusetts to the other, the occasion and progress took on local expression to meet local conditions.

Special Impetus and Response to Factory Organization in Randolph

In Randolph and other Norfolk County and Plymouth County boot and shoe towns, immense orders with big profits stirred and pushed the boot and shoe industrial organization to its very limit of production. Forty-niners who had left shoemaking interests at home naturally saw the chances for marketing Massachusetts boots in California. The same held true in Australia. Randolph boot manufacturers [1] were among the firms to profit earliest by this trade. Mr. Jonathan Wales left his home town in time to have a shoe store established in San Francisco in 1851. At first he bought his goods of the Wentworth, the Whitcomb, and the Howard and French factories of Randolph. His business increased until he was handling boots and shoes from over twenty firms in Randolph, Stoughton, and North Bridgewater. Meanwhile he had entered the firm of Newhall and Gregory, of San Francisco, who sold at auction all the boots and shoes consigned to Jonathan Wales. Mr. Frank Maguire established himself as agent in Melbourne, Australia, and marketed almost the entire output of the Burrell and Maguire factory in Randolph. Both Mr. Wales and Mr. Maguire, as distributing agents, sent home to their employing manufacturers new orders that then seemed incredibly large, and either drafts on Boston banks or bags of gold in payment.[2]

---

[1] The largest boot and shoe firms in Randolph during the 50's were Whitcomb's, Wentworth's, Howard and French's, Strong's, H. Bingley Alden's, Burrell and Maguire's, Clarke's, and Howard's. Of these, Alden's firm had a store in Baltimore and was working mainly on high class boots for the Southern trade.

[2] I have seen at the home of Mr. Wales's son and daughter, some of these bags, made of white drilling and stencilled with large figures giving the amounts, sometimes as high as $20,000. A large silver salver, presented to Mr. Jonathan Wales in 1856, gives the names as donors of twenty men who were connected with the firms which had profited by his skill and efforts.

Adaptation to Needs of Speed and Large Scale Production

Such orders could not be satisfactorily filled with the old rate of speed and equipment. When a hundred or more cases were engaged to go on a certain steamer, due to leave Boston for its route round "the Horn" on a certain day, duly advertised, the boots must be done on time or risk being rejected at their tardy arrival in California or Australia. Formerly the domestic worker had enjoyed all the latitude he needed or wished. He sowed his fields and cut his hay when he was ready; he locked up his "ten-footer" and went fishing when he pleased, or sat in his kitchen reading when it was too cold to work in his little shop.

Manufacturers could never be sure when washouts or snow storms might make the return of stock, sent out to a radius of over twenty miles,[1] well-nigh impossible for days, if not weeks. Account books show that the women who "fitted" and the men who "made" boots and shoes were irregularly employed before the late 50's, and obviously at their own request. They knew that the work must be inspected at the central shop before they were paid, but the pay "would keep."

The rush, then, of the Californian and Australian trade put a stop to this condition in every shop when it brought orders from new trade or increased those from the old. More crimping and treeing, as well as cutting, pasting, and finishing was done "under the roof."[2] There was now a greater inducement to manufacturers to invest in sewing machines for the uppers, and to have the siding up and binding done in the shop by men[3] on

---

[1] Occasionally goods were sent much farther. Nantucket shoe manufacturers rowed over to the mainland at Hyannis and took cases of shoes or boots to bottom that had been sent down from Weymouth or Abington factories. Until the ice on Vineyard Sound broke up towards spring, the finished goods could not be returned by these domestic workers. Cf. story of Asa Jones, p. 137, in Chap. VI.

[2] Girls went into shops to paste, in the 50's. At Burrell and Maguire's, there were twenty-five or thirty girls who pasted in straps and put pieces of bright colored morocco across the upper parts of the vamps. At Whitcomb's and at Strong's girls went into the factory long before power machines, but it was not until after 1865 that they were employed instead of men to run sewing machines in factories in Randolph and neighboring towns.

[3] Cf. pp. 93, 94 for details of machines.

dry thread and wax thread machines, instead of by the women at home. In the previous chapter, figures have been given for the work and wages of men working for Howard and French in their factory in the late 50's on the "Californians." At Whitcomb's Factory, still called a shop [1] in those days, there were eight or nine cutters at work on stock. The whole output of this firm was sold to Jonathan Wales for several years and so was most of the output from Wentworth's. Strong's factory either failed to get a share in this California trade, or did not want it. Mr. Strong sold his product from his Boston office and let Mr. Sidney French manage the factory. His interest was for some time in developing the Congress shoe, already tried out in Weymouth and in Lynn. His was the first Randolph factory to put in the McKay machine, and Mr. French as manager seems to have been progressive enough to have produced up to the standard demanded by Jonathan Wales, and by Frank Maguire. The Burrell and Maguire factory ran to its limit of space and speed, and seems to have made a monopoly of the Australian trade — helped thereto by the fact that Mr. George Maguire, the son, was their sole representative in Australia and had been appointed American Consul at Melbourne.[2]

Transportation in the 50's

The "far away feeling" for not only California, but Australia, had been done away with in many minds in the shoe towns of Massachusetts by the story of "Two Years Before the Mast" [3] which was passed from hand to hand in the 40's and 50's. The shoemakers not only enjoyed the adventure, but had a near interest born of a feeling they were working perhaps on some of the very hides that had come back around "the Horn" as return cargo in just such ships as the *Pilgrim*. The sight of posters and

---

[1] Even now shoeworkers who are over 40 years of age commonly speak of their respective factories as shops.

[2] In 1857, there was a Boston office, as this advertisement shows:
James Maguire and Co.,
Manufacturers of Boots. Factory Main St., Randolph
Store 105 Pearl Street, Boston.

[3] Two Years before the Mast, by Richard Henry Dana, Jr. Published 1840.

the distribution of small colored cards,[1] advertising the sailing and services of ships, added zest to their interest. Just as men and boys know the record and makes of automobiles today, men and boys knew then about various clippers.

The expansion and adaptation of transportation on land was just as rapid and marvellous as that at sea. Railroads were linking up New England and the Atlantic Seaboard States, and running out through the Alleghenies towards the Middle West. Then dispatch companies took up the task of conveyance for the rest of the way across the plains, linking up regions not yet penetrated by the railroads.

Banking Facilities in the 50's

If the tens of thousands of gold could be transmitted safely and promptly from one end of the continent to the other, drafts and checks in mails could be. A larger use of banking facilities and of credit was becoming general. This is illustrated by two pages [2] taken at random from the account books of the firm of C. and A. H. Gilmore, of Raynham, Massachusetts.

---

[1] "A small Celebrated A 1 Extreme Clipper Ship for San Francisco. Merchants' Express Line of Clipper Ships" was the headline of one card. Under this was a colored picture of the ocean, a full-rigged clipper, with sails set, a lighthouse in mid-ocean; "Grace Darling" in large letters, and the picture of this heroine in a red robe against a green background, bound with heavy rope and crossed oars were placed beneath. Then returning to appeal in printed words, there appeared:

Shippers are invited to examine this widely celebrated line.
   Randolph M. Cooley, Manager.
   Agents in San Francisco   88 Wall Street
   Messrs. DeWitt, Kittle & Co.   N. Y.

The Winsor Line advertised in cards so attractive then that they have been preserved among family treasures until today. Here are the contents of one of them:

- Winsor's Regular Line, For San Francisco. From India Wharf.

To sail June 25th. The Celebrated Extreme New York Clipper Ship Golden City, Capt. Leary. This Ship has been put in the most perfect order, and is now in her berth with a part of her cargo on board. She is THE great favorite with New York shippers.

Her passages have been made in 102 and 106 days, while other clipper ships, sailing at the same time, were 120 days.

Her ventilation is perfect, and her cargoes have been delivered in the most perfect order. Shippers who wish to be sure of their goods reaching California in October, will appreciate her.

The well known favorite medium ship CROMWELL, Capt. Adams, now in her berth, and loading, will sail about the 1st of July.

For Freight or Passage, apply to
   NATH'L WINSLOW, JR. & CO.
      corner of State and Broad Sts.
June 7, 1858.   Watson's Press, 25 Doane St.

[2] See Appendix XXIII.

Financial Crisis of 1857 Affected the Boot and Shoe Trade

Bills were not only more promptly paid by wholesale customers on the one hand, but by manufacturers for stock and for labor on the other. Good times led manufacturers to put up larger shops; to install expensive machinery; to put in and cut up more stock in hopes of even better times ahead, just as it led their workmen to build houses on mortgages which seemed reasonable in view of future wages as good and steady as the present. Meanwhile, manufacturers and employees alike in other industries were "booming and spending." As in 1837, so in 1857, overproduction, too great speculation, too much tying up of capital in fixed improvements or investments led to a financial crisis.

Crisis Feared, though not Experienced, in the Randolph Shoe Trade

The general features of this crisis, the so-called Panic of 1857, are familiar to all. A file of the *Randolph Transcript* published that year, quoting news and impressions from other shoe centres, gives a bit of local light and color on the question, at least as it was viewed in 1857, which is valuable for this study. Randolph as a boot centre, closely tied to San Francisco and to Melbourne, was, after all, just a little town doing a relatively large business in boot manufacture. It dreaded hard times, not in general, but in the shoe trade. The first issue of this paper, on March 14, 1857, was congratulatory and optimistic in its general tenor. No hint of depression appears at first. "Randolph has a population of five or six thousand; enough to make two cities out West. That being the case, it seems as though there might be business enough for one printing office."

Boston and Randolph, though only fourteen miles apart, were not then linked by railroad. Stage coach and expresses were transporting passengers and freight. The various boot and shoe firms had their own express wagons or employed private expresses as a rule. One, Cole's,[1] was for the general public to use.

---

[1] "Cole's Randolph and Boston Express leaves Randolph daily at 8 o'clock A.M. Offices at the store of R. W. Turner and at the residence of the proprietor. Returning, leaves office at 34 Dock Square, Boston, at 3 o'clock P.M." Turner's Store was on

This same issue of the newspaper gave a history of the boot and shoe trade of the town to date, closing with the figures taken from the Massachusetts Census of 1855, showing that Randolph had made in that year

345,100 pairs of boots } To the value of
363,300 pairs of shoes } $1,269,400
Males employed for this 1110
Females employed for this 422   1532[1]

"All the other industrial pursuits of the town," it continued, "of such character as to find their way into statistical documents,[2] do not amount to a sum at all comparable with that of the above manufacture," totalling only $129,483. With the valuation of the town amounting to $2,123,440, the Town Meeting of March 30, 1857 had voted $4200 for the schools, $2000 for the highways, and $8000 for town expenses. To keep up this scale of expenditures, the boot and shoe business as the main industry must succeed. The prices of leather [3] and the reports of shoe [4] trade in other

---

"the Hill," the old precinct centre, where the roads from Braintree to Bridgewater meet the roads from the Weymouths and Abingtons, and was therefore central.

[1] This total was practically all the working population of men and women outside the farm and farmhouse.

[2] Randolph's Financial Statistics for 1857:

Hay is raised to the amount of	$23,928
Wood sold or prepared for market	17,050
Boxes made	13,000
Lumber for market	7,700
Potatoes	8,076
Corn	2,005
Rye	157
Apples	2,162
Pears	600
Cranberries	2,650
Live stock	52,155 $129,483
Value of the town	$2,123,440

[3] The issue of March 28, 1857, quotes leather prices:

Sole, light	32 @ 34
Sole, middle	32 @ 34
Sole, heavy	30 @ 31
Calf skin currie	70 @ 80
Calf light	65 @ 75
Sheep skins	25 @ 34

"Stock is small and demand active."

[4] On April 11, this paper announced that "the shoe trade of New York is growing better — trade is increasing both locally and from the South." It spoke also at

states were of great importance. While some days' news must have sounded hopeful to Randolph readers, a quotation from the *Newburyport Herald* [1] must have chilled their hopes:

> The shoe business has been prostrated by adverse circumstances till it is at the lowest point that has been reached for years and many manufacturers especially of women's shoes are reducing the rates of labor and discharging hands. It is found also that the *manufacture has exceeded the demand* notwithstanding the short work.

Conditions in the Lynn Boot and Shoe Centre

About the middle of April, the *Newburyport Herald* reported the failure of Baker and Brothers, "extensive shoe manufacturers of Lynn." It was a bad failure of from $100,000 to $200,000, and the first heavy failure among Lynn manufacturers for ten years, "as indeed now there is the heaviest pressure upon the trade for ten years. This firm employed five hundred hands. Their debts were in Boston and other cities, the heaviest being in Baltimore." The account closed with the announcement that "very many of the manufacturers of shoes are curtailing their business." This was, of course, in the Lynn-Haverhill region. As an editorial, with this news from Newburyport in mind, there appeared in the *Randolph Transcript* an article on the shoe business which gives the best key to one specific occasion of trouble in the shoe manufacture of Massachusetts in 1857:

> It is probably safe to say that not more than half of the towns in New England know about this interest as a manufacture within their borders. . . . In some sections it is the ruling business. . . . In many a town, however, a case of shoes looks as foreign almost as a box of oranges, yet the trade is an important one for Boston, where manufacturers and buyers meet. It is claimed that bad management and want of foresight on the part of manufacturers has caused *leather to be cut up beyond the demand* and thus so great a demand for stock (*i. e.*, uncut hides) has arisen that its *price is enhanced out of all proportion* with that which will be received for it after it is made up into

length of direct trade to begin between Chicago and Europe, quoting from the Chicago Press: "We hear of a sale which transpired on Saturday of 2,000 hides, heavy and kip, for Liverpool by a vessel to arrive here early in May. Also samples of hides are being put up for Scotland to go out by one of the first vessels to leave this port."

[1] Newburyport in the Essex County shoe region was one of the active seaport shoe manufacturing towns.

boots and shoes. . . . It is said that a great *variety of styles of shoes* are in the market, many of which are very slow in going off. . . . It is not unlikely that during this dull time, many workmen will change their occupation, fewer will learn the business [1] and manufacturers will themselves feel freer to curtail work than they do when an oversupply of workmen are dependent upon their business for support. . . . The . . . "good time coming" we hope will be speedy in its approach. . . . The *Southern Trade* which it is said has not been the readiest pay is now the best, while *Western traders* from overdealing have become slow and unreliable. Dealers [2] who were to get high commissions have crowded upon the West a *large stock on long credits*, and thus induced merchants to go "beyond their depth." As we said before, these are only temporary [3] evils, a different class of business men will soon come up and the trade will assume its wonted vigor.

The curtailing in shoe shops did come about, and a week later, April 25, 1857, the *Lynn Reporter* said that many girls who had worked on machines in the shoe shops of that city were now out of employment. "Only five are employed in one establishment, which, until now has had thirty. The manufacturers are curtailing their business." As in all "hard times" a suspicion was cast upon the medium of financial operations. An article which appeared on May 30th on the "Banking System and How Banks do the Mischief," said that "the promise to pay on demand by specie is all sham." Two weeks later there came the "Answer to Specie's Arraignment" in May — which was highly favorable to paper bills.

Conditions in the Boston Shoe Trade

Before June was over, there were various ideas expressed about the future price of leather. It was quoted at thirty cents per pound. Buyers were advised by some to wait until it dropped to twenty cents per pound; yet others thought it would rise after the first of July. "Others in Randolph say no." Some good news came in July 4, 1857.[4]

[1] This is just what happened in the recuperating years of the shoe business after the hard times of 1837, though probably this writer did not know it and thought he was merely prophesying.

[2] This is an interesting concrete statement as to the way in which "over production for markets" is brought about.

[3] Note consistent optimism of this editor in a matter so vitally important to his shoe town.

[4] In the same issue was the news that a "factory burned on June 17th at Canton on the Stoughton turnpike at loss of $1600." Note use of word "factory" instead of manufactory or shop.

Boot and shoe trade in Boston is a little more active. Buyers are in from the South and West for small purchases. Old goods made from high cost stock sell at a sacrifice, while some made from stock at present prices, are slightly remunerative.... The trade in New York City is dull and declining.

Boston steadied her shoe market gradually. On July 11th, it was reported as better still, though buyers were just examining stock, waiting about heavy purchases.

The quantity cleared at the Boston Custom House has been 2596 cases for the week, and 98,723 cases since January 1, showing a falling off from the Boston shipping list of 14,504 cases compared with last year.

Randolph, however, had to face more definite news of the shoe trade which would affect her directly, in August, when, on the twelfth, the Steamer *John L. Stephens*, from San Francisco, reported that those markets were depressed and it was said then that it would need a suspension of shipments for two or three months to relieve them. This struck home to the hearts of more boot and shoe firms in Randolph than to those of any town in Massachusetts, and affected the west village more than the east. In the latter, which was to become a real boot town in the seventies and eighties, the firms still clung to manufacture of shoes, and for a greater variety of markets than was the policy of the west village, *i. e.*, Randolph Centre, which had gone so fully into the Californian and Australian boot trade.

Even with the glut in the San Francisco market, 1002 cases were shipped to California in the week ending August 26, as reported in the *New York Evening Post*, which said that southern buyers were purchasing in the shoe market cautiously. Hides had declined two and a half per cent, and speculators were buying up hemlock soles at the former prices. The western buyers were still waiting to see how crops and payments might be in the West, but the *Boston Shipping List* showed a continuous active demand. About one-quarter as many cases were shipped from New York City as from Boston for California that week. The market was being sparingly supplied by the manufacturers. Evidently the supply and demand were equalizing themselves. The *Randolph Transcript* for September 12th reported that local factories were still running, and the firms were hopeful.[1]

[1] In an article, "Business in Other Places."

108 ORGANIZATION OF BOOT AND SHOE INDUSTRY

As for Randolph, it is much as it has been, though the increasing quantity of leather freighted over the roads shows perhaps that there is going to be more activity in boot and shoe manufacturing. We hope it may be so. Very few if any shops have entirely suspended work. There have been no failures with a single exception,[1] while in many of the manufactories the hands have had no more leisure time than they needed for summer recreation.[2] This is due probably to the choice in the variety of markets offered, and the fact that Randolph firms are not making for any one exclusively, some working for foreign, some for home markets.

Nearness to Boston made hope contagious evidently. The same paper quoted interesting boot and shoe figures.

Boston is now the largest shoe market in the world, and her sales exceed by millions of dollars those of any other city on the globe. There are 218 wholesale and jobbing boot, shoe and leather dealers in Boston[3] whose gross sales amounted to $34,100,000
106 hide and leather dealers........................... 26,650,000
additional sales of retailers............................ 1,390,000
Of the shoe houses, 4 do a business of over........ $1,000,000 annually
" " " " 2 " " " " " .... 800,000 "
" " " " 9 " " " " " .... 500,000 "
" " " " 38 " " " " " .... 200,000 "

The whole number of persons employed in the manufacture of boots, shoes, and leather in the State of Massachusetts was given as 80,000 people.[4] Those employed in boot and shoemaking numbered about 32,000. Of these workers, Randolph was supplying over 1000.[5] The hard times of 1857 left this town of five and

[1] This was the failure of H. Bingley Alden's firm and did not affect the town very materially. It was due to slow pay in the Baltimore markets and came gradually.

[2] This may have been a case of reassurance, or bravado, intended for other towns' ears, or else it may have been the real truth. Tradition in the town confirms the latter.

[3] The United States Census of 1860 gives slightly different figures for Boston. "Boston in 1856 had 200 wholesale and jobbing houses, and the domestic and foreign shoe trade of Massachusetts amounted to $50,000,000." These figures do not include hides, evidently: "The shipments from Boston to San Francisco alone in 1856 were 42,258 cases, valued at $2,100,000." We have figures for New York City two years later in the same Census, "New York in 1858 had 56 wholesale and 600 to 800 retail boot and shoe houses. The sales of

these wholesale houses were......................... $15,000,000
these retail houses were............................. 5,000,000

[4] This was probably from the Massachusetts Census of 1855. I doubt if any later computation had been made and published before September, 1857.

[5] See figures of Census of 1860.

FACTORY STAGE 109

one-half thousand inhabitants [1] in good condition financially, and industrially. On October 10, 1857, the Randolph Bank paid a dividend of five per cent, and the statement was made in print that the bank had a clear surplus of $47,016.72.[2]

The general business depression of 1857 seems then to have done little damage to Randolph's boot and shoe trade, and not to have retarded the development of its factory organization at all. This was true also of Lynn, where, in spite of the newspaper reports of some curtailing of output and reduction of help, with an occasional failure, the hard times made the manufacturers more ready to adopt the Factory System.

Recuperation of the Shoe Industry and Trade from Hard Times and Adaptation to War Conditions, 1857–1865

Recuperation depended mainly upon the success of individual firms in securing better hold in old markets through more efficient organization of their factories. Competition among sellers to win the orders of the southern and western buyers, who were reported as reluctant and slow in placing orders, naturally led to emphasis on good workmanship and speed. Stock was higher in price, and waste could not be allowed. Supervision in the factory would help to save waste and to secure better work as well as speed. Manufacturers realized that the shoe machines, which competitors were adopting because inventors claimed they saved labor, could not be neglected by them. Machines for the uppers, both the skivers and the dry and wax thread sewing machines, were already in very general use, run by men in the factories. Only the lasting and bottoming were done by hand by domestic workers, and even they often worked in gangs or teams in a hired room in the factory, instead of in their own little shops. Times

[1] In 1850, Randolph's population was 4638. It had increased 19.40 per cent in five years, to be 5538 in 1855.

[2] This, of course, may and may not have been a true index to actual financial conditions. During that year, however, the east village (which later became the town of Holbrook) had built a church costing $20,000. The parish was aided substantially in this by Everett Holbrook, one of its wealthiest shoe manufacturers. In May 4, 1857, the newspaper had said that Randolph (including both villages) had several citizens worth over $100,000, and one worth over $300,000.

were good for these lasters[1] and bottomers for they had all the work they could do to equal the speed of the output aided by machinery on the uppers. The same condition prevailed in all the boot and shoe centres, which awaited, as it were, unconsciously, the coming of the McKay machine, which would equalize the speed of work in bottoming and in upper making.

Development of Factory Buildings for the Boot and Shoe Industry

Capital enough had been saved during the good times of the early 50's to build good substantial buildings for the larger stock and output of shoe concerns, and to provide room for both machines and the workmen to run them. Of all the factories in Massachusetts, the so-called "finest" was built by the firm of C. and H. T. Gilmore, in Raynham,[2] in 1857. It was very different from the usual grocery store or private house, turned into a central shop. It seemed like a visible proof of large trade and profits,[3] for it had a wide porch with Doric columns, an office fitted with Venetian shades, and black walnut desks and railings. It was provided with an elevator and dumb waiters, speaking tubes for every room to the office, a steam whistle to summon employees, and a driveway through the lower floor for ease and expedition in receiving and shipping freight. There was a carpenter shop and a special carpenter. The large machine

---

[1] Up to the 60's, a central shop that turned out fifty to seventy-five cases a week was doing a "big business." The Australian trade made the Randolph shop of Burrell and Maguire enlarge its old output of the 40's and early 50's. An offer of $7.00 was made for every extra case over 100 that could be gotten out in a week. One week 112 cases were made in response.

[2] Twenty miles south of Randolph, and three from Taunton, in Bristol County.

[3] The capital for this investment was the result of savings from a shoe business dating back to early in the century, when Mr. Cassander Gilmore, Sr., had a little shoe shop made out of the wash house by the brook, for his Central Shop. He was not a shoemaker but an entrepreneur who saw a chance of profit as merchant capitalist. A few years later he had built a new central shop, some 50 × 100 feet. It was there that the Gilmore firm worked up the trade in children's shoes of fancy colors and men's and boys' brogans and plow shoes for agents to dispose of in Cuba and New Orleans. Such trade in turn, from product made in the old shop, made it possible for the Gilmores to lath and plaster every room in the new factory and have every mop board and door painted or grained.

shop, with a forge and a machinist permanently in charge, was probably one of the chief reasons why Mr. Blake [1] and Mr. McKay came to this factory, five years later, to try out the McKay machines, while Mr. McKay was urging the Gilmore Brothers to help finance the movement of putting the machine on the market. Mr. Cassander Gilmore's son, Othniel, was inventing machines in a little shop in the yard of the homestead across the street. He made the automatic leveller, a press for the binding after it was stitched on the upper, a punching machine for the holes in the quarters of the Balmorals, and a machine to dink the holes in the heel before it was nailed by hand. Although neither Mr. Gilmore of the firm, nor his son, believed very much in Lyman Blake, they entertained both Mr. McKay and Mr. Blake at their home and allowed them weeks of experimenting with their machines in this factory, without installing any permanently.

Perhaps the Gilmores had already put too much fixed capital into the factory itself, to allow them to invest in uncertain stock or rights in the Blake-McKay just then; perhaps they realized that none of the particular kinds of shoes they were making needed McKay stitched soles. Their firm was making for Cuban Spaniards two-holed sailor-ties which had white soles with black and yellow stripes on them "just to catch the Spaniard's eye." The uppers were of black split leather. The plow shoes, which the Gilmores made for the West, were solid vamps and had to be crimped like a boot, but had goring on the side like the Congress shoe. Otherwise it was unlike the Congress or Quincy shoe, for it was of heavier stock and had no fancy stitching. The firm was

[1] In the Superintendent and Foreman of October 27, 1896, the writer of an article on Lyman R. Blake said: "The story of Lyman R. Blake, the inventor, when written by one who is able to write it as it should be written, will read like a powerful romance. With almost no experience as a machinist, he was yet a wonderful mechanic. In Mr. Blake's first machine the horn was stationary and would not sew around the heel and toe; these parts were nailed, but before Blake met McKay he had constructed a second machine, in which the whirl was operated by a beveled gear, practically the same as today. It is one of the remarkable things in mechanical invention, that a machine which finally proved to be of so much value to the world was practically perfect in every essential when it came from the hand of the original inventor." See Appendix XXVI.

also doing a large trade in Balmorals, or Bals, for both Southern and Western markets. This was the output of the Gilmore factory which was the sort of building that seemed to mark the coming of the Factory System into the boot and shoe industry.

New Lynn Factories and their Output in 1860

Lynn[1] shoe manufacturers had also been active in building substantial new factories of brick or wood. A ten-hour system had gone into effect and bells ringing at 6.00 P.M. put an end to the working day. By the returns of the United States Census of 1860, Lynn was manufacturing boots and shoes to a higher value ($4,750,000) than any city in the United States except Philadelphia, whose output a year was valued at $5,500,000. Haverhill was a close third with a value of $4,000,000. These three cities, together with New York City, made one-fifth of the total value of shoes and boots in the United States in 1860. This total was given in the Report on Boot and Shoe Manufacture[2] as $91,891,498. To produce boots and shoes to this value 123,029[3] persons were engaged in different parts of the United States, working in 12,487 establishments, and using a capital of $23,358,527. They consumed that year raw materials worth $42,729,649. This total product of the boot and shoe industry in the United States was seventy per cent above that of 1850. New England's share in the number of establishments[4] devoted to this shoe industry was 2439.[5]

---

[1] Lynn adopted a city form of government in 1850. Her population was then 14,257. In 1852, Swampscott, and in 1853, Nahant, were set off as separate towns, and yet there had been a large enough increase in population in the decade to make Lynn's population in 1860 number 19,083. In 1875, the population had grown to 32,600; in the two decades from 1855 to 1875, thirty-eight new shoe factories were built. [2] U. S. Census of 1860, p. lxviii.

[3] More than one-twelfth of the operatives engaged in all manufactures. Only in agriculture were more hands employed.

[4] Most of these were probably Central Shops. Only a few of them could be termed factories in the modern sense.

[5] The Middle West states had 5412
    The Western   "   " 3175
    The Southern "   " 1365
    The Pacific   "   " 96
Cf. U. S. Census of 1850, p. lxxiii.

New England Boot and Shoe Manufacture in 1860

The New England boot and shoe shops and factories gave employment to nearly $11,000,000 of capital out of the twenty-three millions invested in the shoe industry throughout the United States. They employed 52,010 males, and 22,282 females, out of the total of 123,029 employed in the whole United States in boot and shoe making. The product of their labor was $54,818,148, or nearly sixty per cent of the whole value of boot and shoe making in the whole country. The average value of boots and shoes made in each of the New England establishments was $22,475 per year.

Massachusetts Boot and Shoe Manufacture in 1860

Of all the 2439 manufactories[1] in New England, more than one-half, *i. e.*, 1354, establishments were in Massachusetts. They employed over $9,000,000 of capital, 43,068 male and 19,215 female hands. The product of their labor was $46,230,529, having increased its value 91.8 per cent since 1850. The average capital of Massachusetts establishments was $6655; the average number of hands 46; the average value of the annual product $34,143. The three Massachusetts counties of Essex, Worcester, and Plymouth, together made boots and shoes to the value of more than one-third the total product of the United States.

Boot and Shoe Manufacture in the Brookfields in 1860

The largest single establishment for the manufacture of boots and shoes in the United States was in Worcester County of Massachusetts. It was owned and run by the Batchellers in North Brookfield, and manufactured to the value of more than $750,000 a year. It was the largest of five factories belonging to the same proprietors, which together made in 1860 more than 1,000,000 pairs of boots and shoes. The figures showing the business of this firm in its infancy (1830–31), have already been given,[2] and those for 1875 will be given later when, a decade and a half of growth having been explained, and its development watched, the figures

[1] Cf. U. S. Census of 1860, p. lxxiii.   [2] Cf. Appendix XVII.

will seem less abnormal than if they were given here so close to the humbler ones of 1860. In the neighboring town of Brookfield Centre, a prosperous though less extensive manufacture of boots and shoes was being carried on in those same decades first by the firm of Kimball and Robinson,[1] and later by Henry E. Twitchell. Their market, like the Gilmore's, was the South and the Middle West. They made heavy brogans for the southern plantations and boots, both russet and black, for firms in St. Louis, Detroit, and other cities east of Chicago. For the distribution of the product in the 60's, we have the firm's order book kept from February, 1861, to November, 1865. For the daily details of their business in the earlier years of the factory organization, we have a ledger covering dates, December 10, 1858 to June 29, 1860. These show that here the transition from the large employ of domestic workers for both uppers and bottoms to factory workers with machines was slower in coming to Brookfield than to the Lynn, Raynham, or Randolph boot and shoe centres, for all through these years there were hundreds of pairs of uppers given out to Brookfield women to bind at home. The business of December 11, 1858, and that of June 27, 1860, are much alike as far as progress in factory organization is concerned.[2] Though separated by one and a half years, the first and last pages of this ledger show little difference in the industrial organization used by the firm, and both might have belonged to an earlier decade. Evidently the same kind of brogans and plow shoes were being bound and bottomed by the same domestic workers who had been employed by the firm of Kimball and Robinson for years. Other pages in their ledgers, taken at random, while they show little or no signs of the Factory System already in use in other manufactories of Massachusetts, suggest a reason why this delay had come. Take, for example, pages of the sales book to show the kinds and prices of boots and shoes manufactured by this firm.[3] Most of them were of the old low-priced, roughly made brogan

[1] This firm began as the Batcheller firm did in 1828. For many years they curried leather brought over from Grafton, a tannery centre near by.
[2] See Appendix XXIV for the Kimball and Robinson papers.
[3] *Ibid.*

type. The higher priced ones (even these were selling at wholesale for less than a dollar), were called boots, probably not even short-legged, for no mention is made of crimping,[1] but what we should call a high shoe today. For the higher priced ones, something more than mere essential processes were necessary, and the pages from December, 1858, to January 22, 1859, have carried almost weekly entries of work done by two men on different processes. Charles Hobbs was stitching and buffing a relatively small number of cases, varying from 10 to 24, at the rate of 65 cents a case for stitching, and 90 cents a case for buffing. Philo Walker was dressing boots, sometimes 44 cases at a time, at the rate of 30 cents a case. Whether these men owned the necessary tools and machines and worked at home, or were hired to work in the shop, as Crocker had been at Raynham, and Fletcher was in Randolph, and Leach at Eddy and Leach's in Middleboro, there is no evidence in the books. One man by the name of J. B. Bellows occurs repeatedly by name as bottoming boots at the rate of $8.50 for 60 pair (one case). On December 20, 1858, he is paid also for $16\frac{1}{2}$ D. work, $24.75. This looks like work in the factory at a day wage instead of piece work, netting him $1.50 a day. No entry appears of pay [2] for cutting leather, either upper or sole leather, so that it would appear that the two members of the firm were cutting the sides of leather bought by them in relatively large lots.[3]

The reasons then for the delay in introducing factory methods, or for the late survival of the domestic system here in Brookfield in the firm of Kimball and Robinson, were (1) that they were making a cheap shoe, of no radically new pattern; (2) the firm itself was an old one, the members used to old ways of doing business;

---

[1] Crimping or forming is a necessary process in making boots proper, whether short or long-legged.

[2] It may be that the entry of Dec. 18, 1858, of $323.00 paid Emmons Twitchell was for services as cutter or as bookkeeper. The latter is more probable, for he was their bookkeeper at a later date.

[3] December 21, 1858

686 ft. insoling were bought @ 18c	$123.48
386¼ ft. Calfskin @ 80	309.20
658 ft. Calfskin @	500.08
4533 ft. sole lea. @ 23	1046.75

(3) they were public spirited and had a strongly paternal feeling [1] for the old shoe workers in their village who would lose work by the introduction of new methods and the new machinery in vogue in other factories and other towns. The firm, however, probably as a concession to some younger men in the office or factory, was trying out a few new processes in connection with higher priced boots which they put on the market occasionally. No signal bell has ever told manufacturers when to relinquish their hold on the old, and to introduce more efficient methods of production. We have seen the feverish eagerness of some, the incomprehensible slowness of other firms of our own day in adopting scientific management.

New blood came into this same factory with a new firm in the early 60's, and even the printed headlines of account books show energy and appreciation of up-to-date methods. Pages were so ruled that time was saved not only in entering stock given out to be bottomed and bound, but also in later reference, a mere glance telling the bookkeeper or firm the whole story. Sample pages are printed in Appendix XXV. They show a transition during 1864, when, though the same old printed headings are used, the processes mentioned, like stitching and binding, are done in the factory, and no longer by domestic workers. By that time the old Robinson and Kimball firm had sold out to Henry E. Twitchell, who had been their bookkeeper just previous to the sale. An order book for 1861-65 gives the record of shipping, which shows not only variety of customers and product, but the wide distribution of the same, and the means of transportation in most cases. In Appendix XXV, several pages are reprinted.

Effect of the Civil War on the Boot and Shoe Manufacture

Though the Twitchell books cover the years of the Civil War, there seems no evidence in them of a decrease in demand for

---

[1] Such a feeling existed as late as 1910 in North Brookfield, when one of the members of the late Batcheller firm pointed out the inscription on the public library of the village with pride. It showed the donors of the building to be two humble citizens, man and wife, who had earned and saved all that money during a long life time of binding and bottoming shoes for that one firm.

slave shoes nor of any work on army boots, nor does such evidence appear for any other Brookfield factory. It was perhaps in eastern Massachusetts that the greatest amount of work was done on army shoes with the most far-reaching results.

In the Randolph Centre, 1861-65

Large orders, requiring many laborers to fill, coming at the same time as the great demand for soldiers taken from the number of men who had been boot and shoe workers, called into the Randolph factories and those of nearby towns younger men and more women than ever before, to take the place of the men who had been stitchers on uppers. The hope of cash payments for these large army orders, and the need of speed, induced a more rapid introduction of the McKay machine for bottoms than could probably have come naturally in peace. The stitch of the McKay machine was not as yet perfected, and a demand for old hand-sewed instead of pegged boots brought out old retired shoe-makers [1] who organized into a sort of gang firm, took a building for a factory and got better prices for their work than they had ever enjoyed.

Abington and Army Shoes

It was as interesting as it was natural that Abington, the home of Blake,[2] the inventor of the McKay machine, should have been the boot and shoe centre which made on McKay machines more than one-half of the army shoes provided for the Northern soldiers during the Civil War. Mr. Seth Bryant,[3] a shoe manufacturer living at Joppa, a part of East Bridgewater, took some samples of suitable army shoes, sewed by the McKay machine, to Washington, to present to Mr. Edwin M. Stanton, the Secretary of War. A contract was awarded to Bryant on condition that he guarantee the sewing,[4] and this he promised to do by stamping his name on

---

[1] "Stitching aloft" was done by this band of old retired shoeworkers in Randolph in the shop of Hiram Alden, Sr. Only older men knew the stitch. Younger ones were used to pegging.   [2] See Appendix XXVI.

[3] Cf. Seth Bryant's Shoe and Leather Trade of the Last Hundred Years, pp. 77-78.

[4] Probably this was the very first machine sewed sole the Secretary of War had ever seen or heard of.

every pair of shoes if the Secretary of War would in turn promise to issue an order that no shoes for army use should be accepted without their manufacturer's name stamped on them. This kept the product up to the standard, not only in Abington, but also in Philadelphia, where most of the other half of the army shoes were made. The contract [1] did not prove entirely profitable financially to the Massachusetts producers for several reasons. In the first place, the government was very slow in making payments and the contractors were forced to sell their vouchers at ten or fifteen per cent discount. In the second place, the stipulation had been made that all army and navy shoes should be made of "oak leather." The price of oak tanned leather was twenty cents per pound higher than the hemlock leather, which was common in the Massachusetts markets. Pennsylvania had oak bark as a natural resource and derived relatively greater benefits from the requirements not only in buying its stock easily and cheaply, but in being the market in which the Massachusetts shoe manufacturers [2] were forced to buy. The supply of oak leather in the whole country proved insufficient before the war was over, and a substitute, called Union leather, treated with an oak stain, was used. In the third place, the United States turned several thousand pairs back on Bryant's hands at the end of the war. This was said to be a loss for which there was no redress.

It is certain, however, that the 300,000 pairs of army shoes had kept shoemakers in the Abington region busily employed and had proved an effective establishment and advertisement of the McKay machine.

Brockton and the Development of the "Good Low-Priced Shoe"

Brockton, as Abington's near neighbor, does not seem to have entered upon the army shoe enterprise nor any large use of the

---

[1] Bryant had a contract at the opening of the first bid in New York and the last bid in Philadelphia.

[2] Bryant paid 60 cents a pound in Philadelphia for the oak leather, and one contract cost him $10,000 more for the oak than he would have had to pay for hemlock.

McKay machine. Its start and reputation for the shoe industry came later,[1] with its creation of a "good low-priced shoe" by Mr. Daniel S. Howard, who "knew just how to put every stitch and peg into a shoe where it would do the most good." He got into touch with Fisher and Baldwin in New York before 1850, when they were looking for a good, though a cheaper shoe than the Philadelphia and Newark firms were manufacturing. They in turn told Bryant, of Abington, that his neighbor in Brockton could "make the best shoe for one dollar of any man in the country." Thereupon Bryant, who had a large Boston wholesale shoe trade, told Daniel S. Howard "to put 100 cents into a shoe and show it to him." Bryant closed a bargain with him immediately for 1000 cases, and as he told in later years, he got the lead in the trade in New York City, selling 5000 to 6000 cases right along of Howard's shoes. By that time, Howard[2] was "making more shoes than all the rest of Brockton manufacturers. That established Brockton's reputation for making good low-priced shoes." By the time that Mr. Howard gave up his business in 1888, Brockton had developed this style of shoe by long strides under the Douglas and the Walkover régime, and in 1890 there were 75 shoe factories in Brockton, including Campello, but that date is not within the limits of this investigation.

Lynn's Shoe Factories, 1865-75

Meanwhile, in Lynn, the introduction of factory methods came more rapidly and steadily than anywhere else. The development of factory organization was followed by the outward sign, *i. e.*, power machinery. The finer product, the greater chance for nice machine work done by women on uppers, had led to the introduction of power machines for uppers, even before the McKay was

[1] Arza Keith and his brother had started shoe manufacture in a small way in 1824, in South Bridgewater, which eventually turned — under his grandson, George E. Keith — into the "Walkover" establishment of today. Colonel Southard, Mark Faxon and the Fields were manufacturing in Bridgewater in a "small plodding way" when Daniel S. Howard began business in 1848.

[2] Bryant visited 59 factories in Brockton in 1884, and reported that their aggregate product amounted to $12,208,332, as against Lynn's aggregate of $25,000,000 for that year.

introduced. Steam power was put into John Wooldredge's [1] shoe factory as motive power as early as 1858, to run a machine for making heels. Johnson says that by 1865, the introduction of steam became general in all the large shoe factories in Lynn, and that the "Revolution in the shoe business" was completed by that year.

The Revolution in the shoe business occurred during the ten years ending 1865. From 1855, or a little later, the workmen began to leave the "little shop" to work in the factories of the manufacturers; and in a few years, vacant shops were seen all over the city, until most of them were transformed into hen houses or coal pens, or were moved and joined to some house to make a snug little kitchen.[2]

By 1880, the division of labor in Lynn factories was so marked that Johnson listed the workers. Probably his lists would agree with those which could have been made by other up-to-date factory superintendents in 1875. They are interesting when compared not only to the possible lists for 1815, but with those of 1915, when the division is so much more minute and the processes so much more numerous.[3]

Sole cutting:
   1. stripper.
   2. sole cutter.
   3. sorter.
   4. tier-up.

Upper stock cutting:
   5. outside cutter.
   6. lining cutter.
   7. trimming cutter.
   8. dier-out (of small parts).

Stitching on uppers:
   9. lining maker.
   10. closer.
   11. seam-rubber.
   12. back-stayer.
   13. front-stayer.
   14. closer-on.
   15. turner.
   16. top-stitcher.
   17. button-hole cutter.
   18. corder.
   19. vamper.
   20. button sewer-on.

Lasting and bottoming:
   21. stock fitter.
   22. laster.
   23. sole layer.
   24. stitcher.
   25. beater-out.
   26. trimmer.

[1] This is the same man who introduced the first sewing machine into a Lynn shoe shop.
[2] Johnson: Sketches of Lynn, p. 341.
[3] *Ibid.*, pp. 344–348, and Appendix I.

Lasting and bottoming (*continued*).
- 27. setter.
- 28. liner.
- 29. nailer.
- 30. shaver.
- 31. buffer.
- 32. burnisher.
- 33. channeller, and sometimes[1] a channel turner.

This list of operators in the average shoe factory suggests the number of machines considered necessary for equipment by the time the McKay Sewing Machine for bottoming was on the market. The question of fixed capital, which used to be a small one compared with that of circulating capital in the shoe industry, now became prominent enough to be puzzling, and likely to be prohibitory to new enterprises, undertaken on such amounts of capital as the wage earning shoemakers had saved and were ready to venture.

The Royalty System on McKay Sewing Machines

Colonel McKay, as he watched the rapid increase in demand for his machine for which competing shoe firms all felt the need, by the late 60's and early 70's, realized that sudden unexpected wealth was possible and probable. With a greater vision than most men of his time, he saw three things clearly: (1) if the manufacturers were stocked up too rapidly with the McKay machine before it could be perfected, that the way would be blocked to selling to them the later, improved product; (2) that energetic but poor would-be entrepreneurs in the shoe business could not hope to start manufacturing with the new cost of soling machines added to that of factory equipment and stock outlay; (3) that a steady smaller revenue for a long stretch of years ahead would be even more beneficial in the long run than sudden large amounts from immediate sales. McKay, therefore, determined not to sell outright any of his output of sewing machines. He made it known that he

---

[1] Johnson added that "Boys usually perform these minor parts. . . . In some of the manufactories, the nailing and shaving can be done by a McKay ' nailer and shaver.' A boy sets the nails, a single stroke of the machine fastens them, and a circular motion of the machine shaves the heel with geometrical exactness at one cut, and in an instant of time. The boys who perform these minor parts gain a nimbleness of manipulation that gives them an expertness hardly possible to be gained by older hands." Johnson's Sketches of Lynn, p. 348.

would lease them to manufacturers on easy terms, at the cost to them of one to five cents for each pair stitched. A numbering device was added to each machine to count the pairs as they were sewed. McKay completed this new invention of a leasing system for machines by selling royalty-stamps to facilitate the payment of the royalties. The revenue of the McKay Machine Company rose to $750,000 a year, it was claimed, and the system lasted until the fundamental patents on the McKay Sewing Machine had expired.

Factories in Brookfield and the Contract System

Even with the introduction of similar factory methods of organization, identical machinery and processes, there was opportunity for individual firms to rival their competitors by unique details or minor systems of organization in the boot and shoe industry. An interesting and unique device in factory organization was the contract system tried by the E. and A. H. Batcheller firm in North Brookfield. Mr. Sumner Holmes, who was their foreman in the bottoming room for many years, tells the experience of this contract system which was "such a good thing" for the various contractors that the firm decided, at the end of two years, to assume once more the risks which led to such large profits, and all the contractors willy nilly were reduced again to foremen of their various departments on salaries.[1] He does not tell whether these men as foremen used the same energy and brain power for the firm that they did for themselves while contractors. After that Sumner Holmes, as foreman, took out for his department large sums [2] each month for the firm's pay roll, which give an idea of the business done by this firm in the 70's. The whole set of figures for the firm's business for 1870 is given in Appendix XXVII, to show its volume and departments. The firm was producing stock to the

---

[1] The cutting department was never put under the contract system even when nearly every other department was.

[2] For example,

August,	1869	$21,172.86
October,	1869	20,868.99
March,	1870	13,519.71
September,	1870	18,232.60

amount of two million dollars in a single year. One department of their factory, *i. e.*, bottoming department, was costing them nearly a quarter of a million for labor. Their freight cost them for one year about fifteen thousand, and teaming five and a half thousand dollars. The account of their dealings with the Boston and Albany Railroad[1] alone gives an idea of the proportions which the E. and A. Batcheller shoe manufacturing business had assumed. The town of North Brookfield bonded themselves for $90,000, and paid it in ten years, for building the spur track from East Brookfield. Of this amount the firm of Batcheller paid $80,000 and individuals in the town put in the other $10,000, nearly all savings. The Boston and Albany Railroad leased the track for ten years at three per cent, but the town still owned it. At the end of the ten years, the town leased it again for a fifty-year period.[2]

In 1870, the firm had a much longer list of concerns from which they bought stock than in 1830 (See Appendix XVII). This list [3] shows the development of stock itself, and of allied industries in the boot and shoe industry by 1875. To organize and administer a business like this, a man needed a different training from that of a shoe manufacturer in the Domestic System days in the first half of the nineteenth century. Mr. Francis Batcheller, who treasures the books of the firm from 1831 to 1875, went into his father's business in 1876, when the factory organization was passing into a second, more complete phase.

Summary of the Development of the Factory System in its First Phase

The manufacturers, to save time and to hold the markets by prompt delivery of large orders in the 50's, completed the movement, well on its way in the 40's, of having shoe-making done

---

[1] See Appendix XXVII.

[2] That firm having gone out of business and the lease holding until 1935 makes this spur track a losing proposition for both freight and passengers to the B. & A. R. R. The passenger service is greatly lessened by rival transportation facilities, and yet by the terms of the lease several trains have to be sent over the tracks each day.

[3] See Appendix XXVII.

124 ORGANIZATION OF BOOT AND SHOE INDUSTRY

under one roof under supervision in order to meet competition and the demands of standardizing. This is the chief characteristic of the Factory Stage of the boot and shoe organization; it had entered the industry gradually and almost unobserved. Large buildings, called "manufactories," and later "factories," more capital, larger supplies of stock, were the more obvious features of the growing boot and shoe trade, even in 1855. In 1860, steam power was being introduced into the larger manufactories, making the hand labor of the domestic worker seem pitifully slow in comparison. By the time the Goodyear Welt machine was put on the market in 1875, even though people at large realized only then that the Factory System had come and the Domestic had gone,[1] one whole phase of the Factory System had already passed. The Factory System, which had come in the late 50's, was the prevailing type of organization during the Civil War. The large orders of shoes for the Union armies, added to the scarcity of labor, caused by the volunteering and drafting of soldiers, were additional important factors in urging the use of machinery in general, and in encouraging the trial of the McKay machine. During the war, the practicability of the McKay machine run by steam power was demonstrated, and it was widely adopted during the late 60's. More specializing came on the part of both shoe workers and manufacturers. Some southern Massachusetts towns made shoes only, others boots. Some towns in the western part of the state

[1] The Massachusetts Census for 1875 gives tables showing the number and wages of women furnished with work to do on boots and shoes at home for the year ending May 1, 1875. A few are given here:

Town	No. of women	Total yearly wage	Yearly average wage of each domestic worker
Amesbury	4	$700.00	$175.00
Lynn	575	82,559.00	143.58+
Haverhill	225	20,207.00	89.80+
Randolph	39	1,750.00	44.87+
Stoughton	2	150.00	75.00
Webster	35	1,200.00	34.28
No. Brookfield	6	690.00	115.00
Braintree	1	300.00	300.00

These figures show that there was a marked difference among these few domestic workers, either in the amount, or regularity, or the quality of their work, and also a wide difference in the completeness with which the Factory System was adopted in these old shoe manufacturing centres.

made only cheap brogans for laborers, while others made a finer grade of shoe to be distributed by New York jobbers. The increased variety of styles within this classification made it necessary to dispose of stock while it was in fashion.

The "expansion tendency" of the decade after the Civil War, led, as it did in the 30's, to over-speculation. Shoe manufacturers put more money into railroad stock and western lands than they could steadily hold there, so that when the hard times of 1873 came, many failures were found in the shoe trade. Thus the first phase of the Factory Stage, like the second phase of the Domestic Stage, closed with a sense of disaster and had again to be followed by a period of recuperation. The history of the boot and shoe industrial organization since 1875 makes the story of the second phase of the Factory Stage, which has been characterized chiefly by an intensive system of production, though in common parlance its chief characteristic has been the use of the Goodyear Welt machine. Competition, which has been not only acute but world-wide, has forced economies and heightened the chance of loss on the ever-increasing variety of styles which the product must take to capture the market. The insistent discovery and use of by-products, the absorption of allied industries by some shoe manufacturing firms, and the greater reliance placed by others on highly specialized allied industries,[1] the immense increase in the size of plants and of the number of employees, have all necessitated the perfecting of the Factory System. The rise of the "labor problem" with the closely contested struggles with organized labor has also especially characterized this period. This central phenomenon, however, together with the other factors of transportation, market organization, and finance, which have so pro-

---

[1] The signs seen on buildings when one approaches a shoe manufacturing centre, showing where heels, rands, counters, welting, findings, patterns, lasts, dies, cartons, and boxes are made, furnish a graphic illustration of the scope of allied industries. These make possible what Marshall calls "external economies" arising from "the concentration of many small businesses of a similar character in particular localities"; the "subsidiary trades" which grow up in the locality of a large shoe industry, "supplying it with many implements and materials, organizing its traffic, and in many ways conducing to the economy of its material." Principles of Economics, pp. 266–271 (6th Edition).

foundly modified industrial organization, are outside the limits of this investigation and must be left for later presentation. The beginnings of such organization of the workers on boots and shoes before 1875 will be traced in the next chapter, for the labor problem is even now, in 1920, one of the most puzzling and elemental factors in the organization of the boot and shoe industry.

CHAPTER VI

THE HUMAN ELEMENT IN THE BOOT AND SHOE INDUSTRY

I. *Shoemakers and some of their European Prototypes, 1630–1875.*
   Workmen as Individual Shoemakers.
   (a) Introduction; general consideration of characteristics and influences.
   (b) High standards of life and thought due to
      1. leisure to think and to read while at work,
      2. contact with all classes of people as customers,
      3. influence of legends and songs about famous early shoemakers, like St. Crispin and St. Hugh.
   (c) Did shoemaking take more brains than other crafts? before introduction of machinery? after that?
   (d) Some individual typical shoemakers, and their life history.

II. *Shoemakers of Massachusetts Organized in the Order of the Knights of St. Crispin.*
   Object and propaganda.
   Name and numbers.
   Rites and vows; dues and duties.
   Meetings; external and internal problems; grievances and strikes; relations with Grand Lodges.
   Effects upon the industry.
   Decline of the order.

III. *Summary of Condition of Shoemakers at the Close of the First Factory Period in 1875.*

EXCEPT for personal anecdotes, stories of professional interests and triumphs, and generalizations made from interviewing over a hundred individual shoemakers whose active days closed in the 80's ôr the 90's, there is little to tell of the shoemakers of Massachusetts singly or in organized groups before 1875. Shoemakers from one end of the state to the other have been not only self-respecting and independent, but satisfied and proud over their work. They had outgrown, as a class in the Old England, the beliefs in the shoemakers' traditions and festive customs, but they were still bound in the New England by the teachings of shoemakers who were accustomed not only to all the essential tools and processes, but also to apprenticeship and master workmen, to journeymen and bagmen, either before they settled this country, or through contemporary papers, books and correspondence received here in the eighteenth century.

Without a knowledge, however, of the outgrown customs, traditions, and vocabulary of the early English shoemakers and cobblers who were the predecessors or contemporaries of our New England shoemakers, it is difficult to know just how much creative or original power the latter had or to account for many of their habits. It was not environment alone, nor frontier and colonial conditions in 1700 or 1800 that influenced shoemakers in Massachusetts as to master and apprentice and journeyman relations; as to bespoke and shop work; as to treatment of "bad ware" and to open market sales. They were repeating the acts and rehearsing the opinions of other shoe workers in England.

From such a book as Sparkes Hall's History of Boots and Shoes, and William Winks' Lives of Illustrious Shoemakers, we can gather facts about types and characters which are counterparts of shoemakers in Massachusetts. There have been absent-minded shoemakers, enterprising shoemakers, well educated, philosophical shoemakers in both the Old World and the New. There have been half-trained shoemakers, "mere cobblers" whose bad ware might hurt the shoe market, and whose lower wages menace those of the skilled workman. There were shoemakers working with gangs in bed chambers in England as well as New England before the entrepreneur could afford to hire a shop. There was the witty shoemaker, the political agitator cobbler, the preacher shoemaker on English benches, as well as in Massachusetts. Though shoemakers found an ally in the poet Coleridge, who said that shoemakers had given to the world a larger number of eminent men than any handicraft, his statement [1] might be offset by the fact that "the sons of Crispin have certainly been a very numerous class" not only in modern shoe centres, but in old England, and that, therefore, they might naturally "figure largely in the lists of famous men."

[1] As a boy studying at Christ's Hospital, Coleridge wished to be apprenticed to the trade of shoemaking. Cf. Winks: Lives of Illustrious Shoemakers, p. 189.

"It is felt that something more is required to account for the remarkable proportion of shoemakers in the roll of men of mark. In addition to this, it must be borne in mind that the reputation of shoemakers does not depend entirely on their most illustrious representatives. They have, *as a class*, a reputation which is quite unique. The followers of 'the gentle craft' have generally stood foremost among artisans as regards intelligence and social influence. Probably no class of workmen could, in these respects, compete with them fifty or a hundred years ago, when education and reading were not so common as they are now. Almost to a man they had some credit for thoughtfulness, shrewdness, logical skill, and debating power; and their knowledge derived from books was admitted to be beyond the average among operatives. They were generally referred to by men of their own social status for the settlement of disputed points in literature, science, politics, or theology. Advocates of political, social, or religious reform, local preachers, Methodist 'class-leaders,' and Sunday-school teachers, were drafted in larger numbers from the fraternity of shoemakers than from any other craft.

"How are we to account for such facts as these? Is there anything in the *occupation* of the shoemaker which is peculiarly favorable to habits of thought and study? It would seem to be so; and yet it would be difficult to show what it is that gives him an advantage over all other workmen. The secret may lie in the fact that he *sits* to his work, and, as a rule, sits *alone*; that his occupation stimulates his mind without wholly occupying and absorbing its powers; that it leaves him free to break off, if he will, at intervals, and glance at the book or make notes on the paper which lies beside him. Such facts as these have been suggested, and not without reason, as helping us to account for the reputation which the sons of Crispin enjoy as an uncommonly clever class of men."[1]

Analysis of Characteristic Traits of Shoemakers

After all is said, the high standard of life and thought of shoemakers [2] in both mediaeval and modern times would seem to lie

---

[1] William Edward Winks: Lives of Illustrious Shoemakers, p. 190.

[2] "Shoemakers" is used here as distinct from modern factory operatives on shoe machinery.

in their opportunities and in the ideals which influenced them. *First* in the opportunity already suggested in having the leisure to think, read, and discuss while they were at work; *secondly*, their contact as custom makers, while journeymen, or apprentices in their master's shop, with people of all classes from rulers and clergymen to poor artisans. Even a princess had to have her shoes tried on by the shoemaker; the busiest governor, the most learned judge, and the most pious parson [1] sat by the bench of the shoemaker in colonial, and later New England, long enough to have his foot measured, to order his boots or shoes, and again to have them tried on. *Thirdly*, the influence on their imagination and actions of the century-old legends [2] about St. Crispinanus and his Brother Crispin, and St. Hugh.

English shoemakers were stimulated also by the tales of shoemaker-heroes on the Continent, like Gabriel Cappelini, the Italian painter-shoemaker; like Francesco Brizzio, of Bologna, who passed from shoemaking at twenty to learning design and engraving, and became an artist; or Jacob Boehme, the German mystic of Silesia, all of whom lived in the sixteenth century. Probably many a journeyman shoemaker has been stirred into an ambition for travel by Hans Sachs, the German poet-shoemaker, born at the close of the fifteenth century, who travelled as journeyman, not only to learn his trade more perfectly, but also to see the world. A stanza of one of their songs, sung always on St. Crispin's Day, assured the shoemakers themselves and informed the public:

> Our Ancestors came of a Royal Descent:
> Crispiana, Crispinus, and Noble St. Hugh,
> Were all Sons of Kings, this is known to be true.[3]

[1] It is said that the Randolph ministers acquired the habit of talking over their sermons and threshing out the theological points at Bump's ten-footer, where some of the keenest theological discussions were common in the 50's. One old shoemaker told me it was better for the minister to get "pointers beforehand" than criticism afterwards. Though this must be taken for what it is worth, it reveals a common belief that theological opinions were considered a rightful part of the shoemaker's life at Bump's shop.

[2] Cf. Appendix III and IV and Appendix XXX, XXXI, and XXXII.

[3] Cf. Appendix XXXIII where this poem is printed in full.

Legends like these in the age when they were believed had a strong influence which was transmitted in a subtle way long after incredulity disposed of the value of the legends themselves. Even today in shoe cities where thousands of shoe factory hands are seen going to and from work, there is a definite impression given of self-respect, confidence, well-being, and a high order of intelligence.[1]

Study of Individual Shoemakers as Types

Perhaps the review of a few individual cases of shoemakers in Massachusetts with some reference to their English counterparts or prototypes would serve as a basis for the above generalizations as to the characteristics and standards of shoemakers. First of all we find Thomas Beard and Isaac Rickerman coming over in 1629, bringing a supply of leather and lasts, as well as their kit, confident that the new community would need shoes. They had evidently been invited to join the party because they could be cobblers as well as farmers. Beyond these initial facts of their coming and being given their board and houseroom at the expense of the colony we have no further news. No other individual shoemakers except Philip Kertland and Edmund Bridges, who came to Lynn soon after 1630; and John Dagyr, who came to Lynn in 1750, and taught the shoemakers in Lynn how to improve in making ladies' shoes, have had their advent from the Old England recorded in history. Of the shoemakers born, raised and trained in Massachusetts, some few have become really famous in public life for their other work.[1]

---

[1] There is a tradition in Massachusetts that it takes a higher order of skill and understanding to operate in a shoe factory than in a woolen mill. I do not know whether this could be maintained or not. I do know that a factory hand in Massachusetts resents being spoken of as a mill hand. There used to be exceptions, however, for in Lucy Larcom's time in the mills of Lowell in the 30's, no keener, brighter minds were ever set to daily toil than Lucy Larcom and her companion operatives in the cotton mills. The details of their book reading, paper editing in spare minutes, and of their work as school teachers after money was saved, are very like the story of the shoemakers on the bench in olden times. Cf. Lucy Larcom's New England Girlhood.

Roger Sherman, a Shoemaker whose Trade Knowledge Helped the Continental Budget in the Revolutionary War

Roger Sherman, as a Massachusetts boy,[1] born in Newton in 1721 but brought up in Stoughton, was apprenticed to a shoemaker and worked at the trade until he was 22 years of age. Meanwhile he "received [2] no other education than the ordinary country schools in Massachusetts at that period afforded, but he was accustomed to sit at his work with a book before him devoting every moment that his eyes could be spared from his shoemaking." In 1743 he gave up shoemaking for various other occupations which allowed him even more leisure for studying law, and by 1754 he was admitted to the law practice of Massachusetts. When he was examining certain army accounts during the Revolutionary War, he found a contract for army shoes which he informed the committee, was "defrauding the public by exhorbitant charges." This claim he proved by specifying the cost of the leather and other materials, and of the workmanship. The minuteness with which this was done, exciting some surprise, he informed the committee that he was by trade a shoemaker, and knew the value of every article.

One does not argue from these facts that his shoemaker's training and career made him the great political leader that he became in later years, but it shows the stamp of man that sometimes sat on the shoemaker's bench and gave personal interest to thousands of other shoemakers who followed his later public career with unusual sympathy. He reminds one of a later shoemaker, Henry Wilson of Natick, of whom many details have been given.[3] In his public services, Roger Sherman makes one recall

---

[1] This Roger Sherman, during his later career as a statesman of power and faithful service, was a citizen of Connecticut, but belonged to the United States rather than to any one state, for he was a signer of the Declaration of Independence; a member of the Continental Congress; a delegate to the Convention which framed the Constitution of the United States, 1787; and a Representative and then a Senator in the Federal Congress.

[2] Cf. Sparks Hall: History of Boots and Shoes and Biographical Sketches, pp. 148–155.

[3] Cf. Ch. IV, pp. 68–71.

Hans Sachs, the German poet-shoemaker, who helped his fellow countrymen in a time of public unrest and danger. As a journeyman shoemaker, he associated with meistersingers everywhere he went, and as a master shoemaker he became a leader of his gild even while he was composing poetry. Having laid the foundation for a popular confidence he could be a great help to Luther later in stirring and directing the simple people of his land.

Roger Sherman was not the only shoemaker in New England, or in Old England, who studied enough at the bench to lay the foundations of a later professional or literary career. Like Samuel Drew [1] a shoemaker metaphysician, their English contemporary, Sherman and other shoemakers in the last quarter of the eighteenth century and the first quarter of the nineteenth were reading many many volumes of real literary, historical, or logical merit. In Brookfield we find that a whole group of shoemakers were active readers.

Brookfield Shoemakers as a Reading Public

The village folk of Brookfield in 1820 were readers generally and had a library that was either a public athenaeum or a circulating library owned as a private venture.[2] The latter seems the more likely for I have the ledger, hand-ruled, but business-like in appearance, kept for the years 1819 to 1828. The books were generally registered merely by numbers, yet occasionally by name, *i. e.*, third volume of U. S. History, or Lafayette's Memoirs [3] or Humphrey Clinker, Vol. I.

Whether the readers were too busy to read often, or too slow at this labor to work rapidly, I do not know, but the books were invariably kept out from two to three months at a time. Mr. Amasa Blanchard took books more regularly and frequently than anyone else in the village, generally returning one at the end of two weeks.

---

[1] See Appendix XXVIII.

[2] Apparently by Skinner and Ward, but I cannot be sure, for the flyleaf of the account book does not tell, although the book itself was given to me with all of that firm's regular account books and is identical in style and handwriting.

[3] This book was in greatest demand in 1825, just after Lafayette's visit to this country.

Mr. Daniel Walker, uncle or great uncle of Francis Walker, the economist, was reading the Power of Sympathy, Vol. I, from January 15 to 22d, and Vol. II from January 22d to February 4th. On that day he took Modern Europe, Vol. I, and next month he read Plutarch.

It would be interesting to know if any of these New England reading shoemakers at Brookfield ever read the works of their English contemporary, Samuel Drew, who made both boots and books.[1] Certainly Mr. Walker would have found a kindred spirit in him.

Paul Hathaway, the Belated Itinerant Shoemaker

Paul Hathaway, the "first shoemaker" of Middleboro, represents an interesting local survival of customs, and the transition from the itinerant shoemaker to the custom-maker in his own shop. His town, settled in 1663, retained until comparatively late its frontier conditions, for even in 1796 people were employing an itinerant cobbler. In 1798 Paul Hathaway decided to make people come to him with their leather and let him save his time. He little suspected, I imagine, that in another Massachusetts town as early as 1644 other shoemakers had prevailed upon the General Court to make all their customers come to them in their shops. His own frontier community of farmers had just reached the point where it could enjoy the luxury of having their shoes made entirely outside the household.

Once again Hathaway seems to represent a transition. This time it was from purely custom-made work in his community to sale work sold in general stores. By 1807, his custom-made shoemaking establishment kept three journeymen and two apprentices[2] busy in the fifteen-by-twenty foot shop in his dooryard, but just then Hathaway gave it up on the excuse that sitting on the bench hurt his stomach. It is a matter of suspicion that he realized either that he was a better farmer than shoemaker, or that he could not compete with the sale shoes which were being put on the market and make as good a living. Immediately upon

---

[1] Cf. Appendix XXVIII for the story of this man, Samuel Drew.

[2] These men made the pegs they used and the women of his family spun the flax thread and made their own wax to stiffen the thread.

his return to working his farm with undivided time and strength, he was known as one of the most progressive farmers in that region, and had raised $124\frac{1}{2}$ bushels of shelled corn on one acre, winning a prize that was put on record.

Josiah Field, the Randolph Bagman

Josiah Field, on the other hand, had been one of seven regular apprentices in a Boston shop. In the War of 1812 he was drafted, but his twin brother went in his place. Josiah came to Randolph and worked on shoes for "a Mr. Faxon, who had a Boston store." Josiah Field used to walk into Boston with the boots in his wallet, made like saddle bags and carried over his shoulder. His son, John Field, has described the way his father used to look as he started off on the fourteen-mile walk through the Blue Hills. He was just one of the many bagmen [1] in the state, but one of the best known in his community.

Samuel White, the Expert Pegger

Samuel White, of Randolph, born into a shoe town in 1831, had a better training than Paul Hathaway, though not a formal apprenticeship like Josiah Field. He did not abandon his trade for farming, for he became an expert pegger and a champion seamer. He has told me of his work as a little boy in his father's shop, where he had his stint to do even before he was old enough to be taught the trade definitely.[2] His father was the farm manager for

---

[1] Cf. Appendix XXIX for story of an English bagman.

[2] Samuel White was a type, one of the hundreds of Massachusetts boys, who learned the shoemaking trade right in his father's shop without a regular indenture of apprenticeship. No laws in the state obliged a shoemaker then to show credentials. One realizes the freedom possible then in contrast with the Old England, where the propaganda of the organized shoe workers and legislation, aimed to keep out of the trade undertrained or self-taught shoemakers, made such credentials necessary if a youth was to become a journeyman shoemaker. Many other shoemakers in their old age — all over seventy years, some over eighty years, and three over ninety years — have told me of their share as children in the family's shoemaking. Mr. Loren Puffer, of North Bridgewater, used to get brogans for his mother to fit. She was a widow and so later he went to live with Ebenezer Tisdale a farmer who made shoes during winter months in a bedroom in the second story of his house. The shoes he taught Puffer to work upon he "took out" from shops in Randolph and the boy was taught to last and peg, never to cut and fit.

Dr. Ebenezer Alden and away from home most of the time, so that Sam's brother Solomon, eighteen years older than he, taught him the shoemaking trade. There were several children to be fed and "Uncle Sam"[1] has told of how he used to cry himself to sleep in his trundle bed in the dark attic, hungry enough to gnaw the wooden bed post, because he had left his share of supper to the younger boys during hard times in 1837. Sam's older sister taught him to sew on uppers, and was such a good teacher and stern mistress that he could seam up faster than any man in Randolph or the vicinity. Often speed contests were held in his Union Street shop. Sam could "make a thread" and seam up a boot in 15 minutes. His record in pegging was 12 pair (a case) in one day of boots with double soles requiring 6/8 pegs. He worked with both hands and that made him quicker than the others.

Isaac Prouty, the Rich "Cobbler-Farmer"

Isaac Prouty, of Spencer, Massachusetts, was another shoemaker brought up on a farm with double duties as a youth, which he still combined in middle life. Traditions linger among shoemakers about his appearance and its results. He was once found, by a buyer from the West, out in the field dressed in a faded coat which was without buttons and tied with a rope. When the buyer hesitated to give his order for 1000 cases of shoes, Mr. Prouty led him from the potato field to the barn, which he was using partly as a storehouse. There he told the buyer to rip open any case and take out a sample; that he already had 1000 cases just like "that one" ready to deliver. Dressed in similar fashion, these men love to tell, Mr. Prouty hurried off from his farm to New York City once suddenly, on hearing of an auction sale of leather. His steady bidding up to large sums attracted the attention of his fellow buyers, and even the auctioneer had open fun at his expense. When all the competitors fell off and considerable stock was knocked down to "our poor cobbler friend here," buyers and sellers alike in that market flocked about him, glad to make the personal acquaintance of the man whose name, given quietly

---

[1] Already quoted on p. 94. As a grand army veteran in his blue military cape he became "Uncle Sam" to the children of Randolph.

to the scoffing auctioneer, was so well known to the business world.

On the way home, his face was not known to a new conductor on the railroad, who ignored Mr. Prouty's request and ran past Spencer to Worcester. There the conductor found the railroad officials too incensed and frightened to explain to him what he later realized, when a special train was sent back to Spencer with Mr. Prouty — that this farmer whose clothes smelt of leather was one of the heaviest stockholders in the railroad.

Asa Jones of Nantucket and the Putting-out System

Asa Jones, who was born in Nantucket in 1829, never became a wealthy manufacturer like Isaac Prouty, but remained a shoemaker, cobbler, shoe-retailer all his life, either from lack of capital, enterprise, or opportunity. He is interesting as the pioneer domestic worker who linked up the Island of Nantucket with the mainland shoe manufacturers.

As a youth, he was sent by his father, a custom shoemaker of Nantucket, to Weymouth, to learn any "new tricks of the trade." On his return in 1852, after two and a half years of experience on sale boots and shoes, he helped his father with custom work on high grade ladies' and children's as well as men's shoes, until William Reed of Abington, whose brother had worked with Jones in Weymouth, made a visit to Nantucket and arranged to "put out work" to Jones. The stock — four or five cases at a time for bottoming — was sent by train from Abington to Hyannis and by boat to the Island. For this work, Asa Jones hired a single room [1] in a building on the main street, and one man to help him. The soling was done leisurely in odd times during the winter, for the stock could not be returned until the ice broke up, making it feasible for the boats to go to the mainland. Other domestic workers [2] in Nantucket began to work on brogans for Kingman

[1] One wonders if this was done so as not to confuse the father's stock or prejudice the customers of the old custom shoemaker.
[2] In the summer of 1915, and again in 1916 and 1917, Asa Jones talked with me in his little cobbler shop back of his son's retail shoe store, at first with great reluctance and evident aversion to curious strangers, until hearing familiar terms, like cabbage box, working on the bench, shoulder stick, and straights, made him treat

and Swift, two shoe manufacturers from North Bridgewater, who started a central shop in Nantucket about 1862.

Three Generations of Leach Shoemakers

In the Leach family, three generations of shoemakers spanned three periods of the boot and shoe industry. Levi, living and working under the Custom Stage, succeeded by his son, George Martin, who lived through Custom and Domestic into the Factory Stage, had a grandson, George Myron Leach, who learned in the Domestic Stage and worked in factories until after 1889. Their story is probably so typical of eastern Massachusetts that it can be used here as an illustration of the trend of the times and trade.

Levi Leach (born 1775 in Halifax, Mass.) was a farmer in South Bridgewater who taught school in winter and made shoes in a ten-footer, which he built in his "side front yard." His work was custom work. Where he himself learned the trade is not remembered, but knowing conditions in the Bridgewater regions during his youth and early manhood, would make us wonder if he had been an itinerant cobbler, like Paul Hathaway, who settled down in his own shop and began custom work.[1] This Levi Leach, besides teaching school three months each year, taught his three sons, George, Levi, and Giles, to make shoes. When the eldest son, George Martin (born 1821) was twenty-one, he went to East Middleboro and repeated his father's program in buying a farm and farming it, building a ten-footer in his orchard near the street, and teaching the village school three months a year. The work in his shop, however, was not of the custom sort. He was soon buying stock, cutting it up, both sole and upper leather, for brogans and Oxfords, and sending the uppers out to women in the neighborhood to bind and side up. After this work was

me like a comrade. He, of course, had no conception of the Domestic, or putting-out system which he had helped to introduce into the island, nor a thought of his having shared in the transition from the custom to the domestic system, though he had a retailer's instinct for thinking "factory made shoes were the only thing for these days," and did not now resent the factory system.

[1] If so, his life experience included the second phase of the Home Period, and he with his descendants spanned all four periods.

brought back to his ten-footer, which now had become a central shop, the uppers were inspected and sent out to be bottomed in various ten-footers in the near neighborhood or by people "down Plymouth way." By the time his younger son, George Myron (born 1845) was seven years old, George Martin Leach had outgrown his ten-footer, and had been taken into partnership by Deacon Eddy [1] of East Middleboro, who already had a successful grocery business and the post office on the lower floor of his two-story square building. Deacon Eddy put into the venture $10,000 which he had made in the shovel business, and Leach put in only $200 in cash, but brought knowledge and experience of the manufacturing of shoes. Their business was largely in brogans for the Southern trade, mainly for New Orleans. The second story of the Eddy building was used for their central shop. Here both sons, George Myron and Giles, learned the trade, pasting linings in brogans, then closing seams. The stint for a twelve-year old for a Saturday forenoon was to side up ten pair of brogans, using barrel staves for clamps. By the time George was old enough to understand machinery, his father's firm had invested in a stitching machine for uppers and George Myron and his brother Giles had full charge of the stitching in a small fourteen by fourteen ell added to the back of the building.

By 1860, when Deacon Eddy was ready to retire from the firm, the Eddy and Leach central shop was becoming a factory where most of the bottoming as well as the cutting and crowning was being done under supervision in the shop. Leach kept up the firm until 1874, but instead of competing with other manufacturers who were buying and installing expensive machinery, he became more and more a jobber, buying up and selling shoes already made. Meanwhile both sons, George Myron and Giles, had moved to Raynham and were working in shoe factories there and in Brockton as stitchers, George keeping it up until 1889.[2]

[1] Cf. p. 71 where this is mentioned.
[2] It was natural that these sons, George and Giles, of the third generation of shoemakers, being born into an age of specialization, made stitching their sole business, and never farmed nor taught school, though each owned a farm, and were proud of their sister Anne who did teach. Of George Myron's four sons and two daughters, none ever worked on shoes but one son, and both daughters and one

Lucy Brown, a Contented Rapid Stitcher

Many other shoeworkers have similar memories. Mrs. Richmond Brown, born in 1840 at Hanson,[1] has told me of going, as a girl, to the central shop in North Bridgewater, to take out boots for her aunt to cord, wheeling them to and from the house in a baby carriage. By 1861 she was stitching by foot power on a wax thread post machine so many moccasins [2] at seventy-five cents a pair that she made four dollars a day easily. If she did not get at least eighty dollars in her monthly pay envelope, she said she was disgusted, but she generally did, for work was plenty and she had all she could do. She said that a man working beside her gave up his job because he could not make more than three dollars and a half a day where she and other young women were making four dollars at the same job. By 1863 the machines, Wheeler and Wilcox, were run by steam power in Reed and Clapp's shop,[3] where she worked through the 60's. Not many men ever worked in the fitting rooms, for there was work enough for them in the lasting and bottoming rooms. Mrs. Brown's work never tired her out; she and her fellow-workers never had any grievances. They were not members of any protective shoe organization and felt no need of any. She always enjoyed her work, and as she told of it at the age of seventy-seven her eyes shone and her hands described the motions unconsciously. When the present Commonwealth Factory began in 1883, she stitched the uppers of the first sample shoes they made, and continued to work there until 1912.

Both she and her husband said repeatedly, in discussing the St. Crispin movement, that they and their friends and most of their neighbors, all shoemakers, in either the cutting, fitting or bottoming departments, had no grievances, no problems, and did not care to organize as laborers. They probably represented the more skilled and highly paid workers of the old English stock of

grandchild have become teachers, making five generations of school teachers, and at least three of shoemakers in the same family.

[1] Formerly a part of Pembroke, Mass.

[2] These moccasins were made of cloth, trimmed with leather, and lined with buffalo skin for men, and with lamb's wool for women, and soled with good leather.

[3] This shop was an old meeting house fixed up into a four-story factory.

Massachusetts. Some of the Irish immigrants of the late 40's and early 50's felt as they did. Some journeymen in Massachusetts, however, thought they had grievances in the late 60's.

Labor vs. Capital Problems before the St. Crispin Organization

It has always been hard for individual journeymen shoemakers, without experience in the manufacturer's problems, to realize that the factory organization brought into the industry entirely new problems, such as having large amounts of fixed capital involved in stock, in machinery and factory buildings, where, in earlier periods, the circulating capital with which to pay wages was far larger proportionately than that invested in central shops or stock. Overhead charges were slight in a ten-footer or a 50 × 100 Central Shop, even when credits were necessarily long, as in the Western and Southern trade. Entrepreneurs came more and more from outside the trade. There is an interesting exception, however, as late as 1881, which is typical of earlier decades. M. A. Packard, of the Abingtons, was docked in the cutting room, as the story runs. He would not "stand for it" and, buying up a stock of leather, cut out and had made up a case of boots, took them to Boston and sold them promptly at a jobbing house. With the larger capital, he made his business wider and wider, until his firm, the M. A Packard, was one of the largest in Brockton. His spirit and the adventure had been common in the 20's and 30's.

In early days, the advance in wages could be passed on to the consumer; but in case of production for widening markets and solicited orders which involved bargaining and planning ahead for wages, the manufacturers could not change wages at will without doing it at personal loss. It was but natural that the workmen should always see the increase of business, in size and number of grades, without appreciating the attendant increasing risks, losses, and problems. Neither did they take into account that they, by their demand for a still larger share of the profits as increased wages, might force the manufacturer to decide to give up the business altogether and invest his capital elsewhere.

These were some of the problems and possibilities in the shoe industry that the St. Crispin organization did not realize sufficiently from 1868 to 1874.

Shoemakers in Massachusetts in Shoe Organizations

After the initial "Company of Shoemakers," the so-called Boston Gild, which was incorporated in 1648 under a charter granted by the Colony of Massachusetts Bay, no attempt at organization of shoemakers as such was ever made or succeeded [1] in Massachusetts until the Knights of St. Crispin, organized there, as in all other shoemaking parts of the Union, in 1868.

In the first case, that of the Boston Gild, the shoemakers appeared before the public with the vicarious object of preventing the making of "bad ware" by inferior workmen. It must have been realized even then that the move was not wholly unselfish, that the "bad ware" must affect the good shoemakers indirectly.[2] No definite prices of goods to the consumer, nor scale of wages of journeymen shoemakers were determined. It was not an attempt at a closed shop, but at gathering into the shop the parts of the work formerly done by the family aided by the more skilled itinerant cobbler. While it advertised to protect the public from poor work by taking the itinerant cobbler into the master's shop where his work could be watched by the gild officers, it subtly opened the way for masters of the trade to make more of the shoes of the community in these same custom shops. We have seen the passing in the city though not yet in the country of the "itinerant"

---

[1] As yet I have found no trace in the Massachusetts shoe industry of the existence of the Society of Master Cordwainers which was in Philadelphia in 1789; of the Federal Society of Journeymen Cordwainers, which was in Philadelphia, 1794–1806, or of the United Beneficial Society of Journeymen Cordwainers organized in Philadelphia in 1835, mentioned and described in the Quarterly Journal of Economics, November, 1909, by Mr. John R. Commons, in an article on the American Shoemakers.

"There was a shoemakers association in Lynn as early as 1651." Article in City of Lynn Semi-Centennial, Lynn Fifty Years a City, p. 66. Printed by direction of Celebration Committee, Lynn, 1900.

I have never found this statement elsewhere. I do not know whether it is correct or not.

[2] Cf. Appendix VI for their charter and the general history of the movement.

dressmaker of our own days and of the unprofessional or undertrained nurse who could not be accepted in an association of skilled women of her profession.

When the Knights of St. Crispin organized two centuries later, they still feared the inferior workman, but not his work. They were determined to oust from their trade "green hands" whose work on machines made as good if not better shoes than the ordinary journeyman shoemaker could turn out by hand processes and tools.[1] This state of affairs had not been reached until the late 60's, when the McKay machine was well established and tested, so that even a Chinaman who understood not a word of the English used by the foreman in the factory, who knew nothing of the processes of shoemaking, could turn out a satisfactory piece of work by using his eyes and imitating motions which he might or might not understand. When the threat of using Chinese labor with none of the boasted Yankee ingenuity was made by manufacturers and carried out by Sampson, of North Adams, it was an overwhelming surprise, a crushing answer to the arguments of Massachusetts boot and shoe workers that shoemakers had to be men of skill and highly intelligent. They had not, as they protested then, and do to this day, objected to machinery itself. As machine after machine was introduced between 1835 and 1865 to do work on stock or to aid processes on uppers or soles, the journeyman shoemakers had not protested for they expected to have chances to work at these machines. While apprentices who developed into full-fledged allround shoemakers became fewer and fewer as young people of a shoe town went into central shops and factories to learn just one process, and the members of the gang of bootmakers were expert at only one part of a process, the journeyman shoemakers of Massachusetts were content. They were busy; they were earning good wages steadily during the

[1] Commons says (In American Shoemakers, Quarterly Journal of Economics for Nov., 1909, p. 75) that the "factory succeeded in producing a quality of work equal or even superior to that produced by the journeyman." This needs qualification to be accepted by real shoemakers of the past or present. The McKay sewed shoe even in 1868 could not be called a superior product to the shoe made by a custom shoemaker of skill and full training. Even today, with the perfected Goodyear welting machines and various machines for turned shoes of high grade, there are whole factories devoted to making by hand shoes which sell for fancy prices.

"runs" of each year. They did not have the grievances felt so bitterly in the Philadelphia centre of having to compete with prison labor and sweatshop workers.[1] In fact they were themselves considered a grievance to the better class shoemakers of Philadelphia. Yet each period of rapid expansion in the shoe industry, especially the one following the War of 1812, the ones that culminated in or were checked by the panics of 1837 and 1857 in turn, had brought into the trade the marginal producer, the under-trained or the unskillful shoeworker. This had been, however, for confessedly poorer work, to sell at a lower price, such as sale work, market work on speculation, or work for Spanish consumption in South America, Mexico or Cuba. Whenever manufacturers were competing against each other for labor, poor workmen had stepped into the trade at relatively good pay. But at the time of a panic or a slump in business these had been the first to go. Each time the recuperation of the shoe industry was characterized by more precise workmanship, more styles, and more standardization. The poorer workmen [2] were the first to be dismissed under such a régime of readjustment. Such inferior workmen could, however, temporarily reduce wages and were always somewhat feared as well as disdained by real shoemakers.

In 1867, however, the green hand, the untrained shoeworker, might be a mere machine operator and beside turning out good work, could usurp the journeyman's place or else reduce his wages. Manufacturers, with hosts of unskilled laborers to call upon, need not compete with each other and send up wages. On the other hand, those manufacturers who did not have machines

[1] In 1835 Philadelphia shoemakers publicly complained that the Eastern States, meaning Massachusetts, did not do shoemaking as well as they, and charged less. This was probably true in all its order and sale work aside from its regular private custom work. New England was specializing then in brogans and cheap shoes for women. The Philadelphia shoe industry always made the highest grade shoes with skilled German workers. Boots were never made save as custom work. Those workers probably felt about Massachusetts shoemakers then as our Lynn and Brockton Union shoemakers feel about the non-union workers in Maine today.

[2] Samuel Cox, of Lynn, has told me of his difficulty in getting work in 1838. He was not a shoemaker, but a workman who had learned in less than a year to do work on cheap children's shoes and had found plenty to do before 1837.

turning out eighty pair where a journeyman worker could do one pair, had to reduce wages to survive at all. Shoe shops, the shoemakers felt, must be closed against such green hands, and manufacturers who would not agree to it, must find empty workrooms. Organizing leaders said all the shoemakers must stand by each other in this movement. Massachusetts shoeworkers, to a large number,[1] joined the country-wide organization of Knights of St. Crispin, helping to swell its total membership to forty or fifty thousand.

Origin and Sources of Information concerning the Knights of St. Crispin

A Massachusetts man born and bred, formerly a bootmaker in Milford, then at work in Wisconsin, Newell Daniels by name, was the leading spirit of the union when it was organized in Milwaukee in the spring of 1867.

The name of the organization had much to help it in its centuries-old usage among cordwainers or shoemakers. The wealth of St. Crispin legends [2] and songs could easily be revived. The history of the original sixteenth-century Order of St. Crispin could be effectively rehearsed in meetings of their nineteenth century brother shoemakers. Their meetings and rites, their initiation vows and pledges were secret, but their demands as laid before the manufacturers and their objects as stated to fellow shoemakers whom they sought to persuade or force into membership were public property then as now. Within the last ten years research students [3] have found and collected important original material on the subject, such as the proceedings of the Grand Lodge, speeches of Grand Knights, reports of officers, besides the text of the Constitution of the Knights of St. Crispin. From such sources, added to popular stories of old shoemakers, a fairly life-

[1] Investigation does not prove, however, that even a majority joined the Crispins.

[2] See Appendix XXX–XXXIV for some of the Crispiniana that had accumulated in literature before 1868.

[3] See Lescohier's use of such materials in "The Knights of St. Crispin 1867–1874." Published in the Economic and Political Science Series of the Bulletins of the University of Wisconsin.

like picture of the workings of the St. Crispin order can be drawn today.

Objects of the St. Crispin Organization

The object of the organization [1] was: (1) to keep out of the shoe trade all new workers, "green hands," except sons of St. Crispin members who might be taught the trade; (2) to keep the prices of work on boots and shoes up to that demanded by the skilled workers; (3) to refuse to work in any shop or for any employer who employed scab labor or who would not keep a "closed shop," in modern parlance. The opposition of the Order of St. Crispin, as stated by the members both then and now, was always to the "abuse" of machinery, *i. e.*, its operation by non-shoemakers, and not to machinery itself. This seems a mere quibble to us in one way, for we are so accustomed to the idea of installing "fool-proof machinery" that will save the employment of skilled labor. But if we realize that the investment of capital on a large scale in machinery was not familiar to them in its intent and results, they seem sincere enough.[2] The object was also (4) to encourage coöperative manufacture which would assure not only fair wages and more regular work but self-employment. This was announced in the Preamble of the Constitution: "We believe also in coöperation as a proper and efficient remedy for many of the evils of the present iniquitous system of wages that concedes to the laborer only so much of his own productions as shall make comfortable living a bare possibility, and places

---

[1] See Appendix XXXVI for text of Article X which states the object as far as "new help" is concerned. Mr. John F. Tobin, General President of the Boot and Shoe Workers Union, wrote in 1915 as follows: "I have never found anyone who could give me any connected history of the Knights of St. Crispin, notwithstanding the fact that I have made diligent search. I was a member of the organization some forty-three years ago, but at that time, did not accumulate any knowledge of its workings.

"I feel that its history was not of any great consequence and its achievements did not go beyond seeking higher wages through the old-fashioned strike method. At that time, employers were practically unanimous in opposition to organization of the workers, and the contests generally hinged upon a recognition of the Union in conjunction with the wage question."

[2] "We liked the machines, but we wanted to have a chance to run them. We'd been born and bred as shoemakers," is the testimony today of old workers who belonged to the Order.

education and social position beyond his reach."[1] This belief in coöperative production was the only constructive propaganda that the order evolved. The main vigor of their order was spent in protective measures as to wages and "green hands."

Membership

Out of a total membership of 50,000[2] claimed in 1870, we do not know how many were Massachusetts men, though the American Workman of March 5, 1870 claimed that there were 40,000 Crispins in that state. There were, however, three local lodges formed in Massachusetts in September of 1867; forty-three formed during 1868; sixty-seven were recorded as being active in the state in 1869, and eighty-five active lodges at the end of 1870, when the order as a whole throughout the country was at its height, in power and enrollment. That Massachusetts should back up Newell Daniels, the founder, was natural. That Lynn and Worcester should have the most members and the severest strikes over grievances seems natural too, since they were the largest shoe centres. Yet Massachusetts had the same falling off in numbers, suffered from the same delinquencies of members and fellow lodges, and was open to the charge of lukewarmness just as much as any other state which joined the National Grand Lodge.

Rites and Vows

There seems to have been no really effective way of keeping members true to their initiation vows; no good system of collecting dues; and no adequate framework of rules to keep a good organization, in spite of the high aims and worthy services of certain founders and leaders of the Order of St. Crispin. Just what the initiation rites were remained a profound secret during the lifetime of the order, though shrewd guesses and waggish stories[3] gave a popular conception. Fortunately there is re-

---

[1] Cf. p. 148 — where mention is made of the difference which came in the attitude of members towards the more prosperous workmen in their order who became foremen and then employers.

[2] See Lescohier: Knights of St. Crispin, pp. 7–8, for discussion of enrollment.

[3] I have had initiation rites described to me by a shoemaker who did not join

corded the vow or initiation oath supposed to be taken by all members which gives such an adequate conception of the promises and beliefs of individual Crispins that it is reprinted here in full.

I do solemnly and sincerely pledge [1] myself, my word and honor as a man, before God and these witnesses present, that I will not divulge any of the secrets of this Lodge to any one who I do not know to be a member in good standing, except my spiritual adviser. I will not make known any of the signs of recognition or any matter pertaining to the good of the order. I faithfully pledge myself, that I will not learn, or cause to be learned, any new hand, any part of the boot and shoe trade, without the consent of this lodge, and I will do all I can to prevent others from doing the same. I shall consider myself bound, if any member shall violate this rule, to be his enemy, and will work against his interest in every way possible without violating the civil law. I further pledge myself that if a member gets discharged from a job of work, because he refuses to learn a new hand, that I will not take his place, except the member discharged gives his consent. I also agree to be governed by the will of the members of the order. This pledge, I agree to keep inviolate, whether I remain a member or not, as long as the organization stands. So help me God.

Before a shoemaker could take this vow or become eligible for membership he had to work for "an aggregate of two years at boot and shoemaking," bring forward evidence satisfactory to the lodge he proposed to join, and be engaged at his task at the time.[2]

If a shoemaker became an agent or a foreman or a manufacturer, even on the smallest scale, he could no longer be a member of the Order. Henceforth his interests were supposed to be allied to capital and hostile to labor.

Dues and Duties

By the terms of the Constitution of the Grand Lodge of the Knights of St. Crispin and by those of local lodges, the dues were fixed [3] at thirty cents as a per capita tax to be levied yearly on

the order. Just how he got his information or whether it was a popular guess I cannot judge. The candidate for membership, he says, was swung to and fro in a blanket held, jerked, and pulled at the ends by two members. When a certain old man by the name of Shannahan, of Abington, was being inducted, tradition says the ritual was
"Welkim, welkim to this band
Here goes up old Shannahan."

[1] This pledge was printed in the "Hide and Leather Interest," issue of May, 1869. Cf. Lescohier, p. 29 and Appendix XXXVI.
[2] See Const. Article XVI on membership. [3] See Const. Article IX on revenue.

each member of subordinate lodges. One half of the amount was to be collected in the month of July; the balance in the month of January in each year. These were the dates when the winter and summer runs were in full swing so that the money could be easily spared and readily given. This amount was to cover the overhead expenses of the organization. Each member of a subordinate lodge was to pay into the local treasury fifty cents as a contingent fund for meeting the expenses of "grievances" *i. e.*, strikes and individual non-employment enforced by obedience to the vows of the Order.

Because many lodges failed to exact and collect this contingent fund, the strikes were badly and tardily financed. Several former Crispin members have told me that there were no dues in their lodges. This statement at first seemed to indicate a lapse of memory on their part, but after reading the indictment made by various officers of the Grand Lodge against local orders, one is forced to believe that the given facts represent actual conditions in some places. Normally each Crispin was supposed to be taxed thirty cents a year and perhaps eighty cents if one strike occurred in his province.

The duties of the lodge members consisted, as their vows indicate, in resisting invasion of their rights as shoemakers. They were expected (1) to refuse to teach a green hand even if they lost their jobs by so doing; (2) to take a share in the meetings and in pushing the propaganda of their order; (3) to hold whatever offices in the lodge they were chosen to fill; (4) to refrain from discussing within the lodge room the merits or demerits of any religious denomination or political party;[1] (5) to refrain from injuring the interests of a brother Crispin by undermining him in price or wages or by any underhand act;[2] (6) to use their utmost endeavor to induce persons who work for them or with them in their trade to join the Order.[3]

[1] See Const. Article XIX.
[2] See Const. Article XVIII.
[3] See Const. Article VI of the subordinate lodges.

External and Internal Problems of the Massachusetts Lodges of the Order

Perhaps the strikes were the most prominent of the external problems as well as the most effective weapons of Crispins. The strike of 1870, lasting for three months among the shoemakers of Worcester, involving 1200 men and costing $175,000, made anxious days and heavy financial obligations for Massachusetts Crispins. The Lynn lodges were so strong and successful in the first strikes of 1869 and 1870 that the manufacturers "hardly dared to take orders." Some firms moved out of the state, to the eventual loss of the Crispins who had thus narrowed the market for their labor. The North Adams strike of 1872 was ended for the Crispins by the unexpected but determined action of Calvin T. Sampson, who brought 107 Chinamen from California for his factory.[1]

The greatest internal problem, a very real and fatal difficulty, was that of financing the work of the Crispin Organization, already mentioned [2] in connection with dues. Lescohier, our chief authority on the history of the Order, says "the Crispins failed even from the first to finance their grievance strikes adequately; and the failure was fatal." The man who gave up his job for the cause could draw at best but six dollars a week for himself, two dollars for a wife or mother dependent upon him, and one dollar for each child under twelve years of age. This aid barely kept them from want in winter and when the local lodges neglected entirely, or in part — through indifference, lack of sympathy or of efficient officials — to collect and pay over moneys due the strikers or grievance bearers, the Crispin order suffered in the eyes of the public and the unprotected members were ready to leave the lodge.

The great distress felt by the leaders of the Grand Lodge over this financial situation was voiced in Boston at the third Annual

---

[1] Mr. Jerome Fletcher, quoted in the opening pages of this book, often told me with glee, from the manufacturers' and the non-Crispin point of view, how the Crispins "reckoned without their host" at that North Adams factory.

[2] Cf. pp. 148–149.

meeting of the Grand Lodge on April 19, 1870, by the Chief Knight of the whole order:

You perhaps think it strange that I dwell so much upon the duties of deputies, but after I explain, you will see the necessity. At our last annual session we had some evils to overcome which were great obstacles to the well-doing and carrying on of the institution. The greatest of these was that some of the lodges were perfectly indifferent with regard to the payment of the International and Grievance taxes. While the Chicago grievance was going on, there were certain persons who discouraged the lodges from sending in their taxes. It is very easy to discourage individuals from paying out money, so easy that one man wishing to get the good will of the majority, can baffle a dozen of good men from sending any money to the International or suffering lodges. I want the deputies to know their duties, and do them. I want them and the lodges to understand that they have their authority, from the International Lodge, and that the lodges must obey and respect them, particularly when they remind them that they are doing those things which ought not to be done, and leaving undone those things which ought to be done. . . .

I sincerely believe that if the deputies had understood their duty, and the lodges had known their power, all the lodges who have had recognized grievances would have received the money due them at the proper time, which has not been the case. Many persons have said to me, "Are you not discouraged?" My answer invariably has been, "No, sir." "Well," they would say, "I am. I think the Order never saw such trying times. We are terribly in debt, and no money to pay it. The plan of keeping the money in the lodges is a failure." And they would go on to tell how the members of the International Lodge did very wrong in adopting such a plan — that they might have known better, etc. Let me say, brothers, it is very easy for us to say that anything is wrong after it has been proved to be so. It is very easy for us to say that a certain measure is wrong after we have tried it for one year, and it has proved to be a failure. But let me say that this idea of having a contingent fund in the subordinate lodges, *is a good one.* I believe that there can be no better plan for the safe keeping of the money, and that it can be sent to any point where a grievance may arise as soon as from two to ten days after the grievance has been recognized. The machine is right, but the operators are wrong. Let me just cite to you briefly the manner in which this department has been managed. When the C. S., of a lodge receives notice of the recognition of a grievance, and that his lodge is called upon to send a certain per cent of their contingent fund in support of it, he puts the communication into his pocket, and forgets all about it, till after the meeting, and if the lodge meets but once a fortnight, it will be in his possession about three weeks before the lodge knows anything about it. He then brings it into the lodge and reads it; it may be acted upon, and it may not; if it is, you will find perhaps a member who goes against everything that is brought up, unless it may be something for his benefit. . . .

Last International session we found ourselves in debt about ($20,000) twenty thousand dollars. The Executive Council cleared up all the apparent mysteries of the Chicago grievance, so that all the delegates were perfectly satisfied. They all promised to go home to their lodges, and if they were in arrears to induce them to send the money right along to pay Chicago; but the money did not come. Why not ? I answer, because the rank and file did not understand it. The delegates did not take pains to push the matter. At the meeting of the Executive Council, held in October, the Council took into consideration the question of delinquent lodges, and came to the conclusion that it would not be just for them to levy a tax upon the lodges that were not in existence at the time of the Chicago grievance, neither did they think that it would be just to ask the lodges who had paid their dollar and twenty-five cent tax, to pay it over again. And neither did they expect that they would pay any assessments for those grievances until the delinquent lodges were obliged to pay. We knew that we had no power to force these lodges to pay their taxes. So, there was nothing left for us to do but to ask each member who had a heart in this institution to contribute voluntarily one dollar or upwards, for the purpose of paying our debts; debts that were contracted in fighting our battles. But there has been very little money received on that call. . . .

Now, I firmly believe if the officers of the lodges had made an effort to collect the money on that call, we would not be in debt one cent today.

Like the Confederation of the United States in 1776 to 1789 where Congress and the Executive Committee of Nine could not force the constituent elements of the organization, the states, to pay the money requisitioned as their quota, so the Grand Lodge of the Knights of St. Crispin could not force the local lodges to pay their just share. There was more hard feeling and actual cause for anxiety on this score than on any other until the end of the Order was admitted to have come in 1874. Long before that another cause for disquiet was felt by leaders and by thinkers in the organization over the divergent views as to members who became prosperous. By the Constitution already quoted [1] membership ceased automatically for a shoemaker who became a foreman or manufacturer, except he was employed under a coöperative system. Even here he was viewed with suspicion by many who had avowed themselves in sympathy with coöperative boot and shoe production. When Mr. Newell Daniels as International Grand Scribe was reporting to the National Body at its meeting in 1870, he attacked this question boldly.[2] As the

---

[1] Cf. Const. Article XVI.

[2] "I wish to call the attention of this body to Article IX, Sec. 3, in regard to eligibility of membership, in relation to foremen and manufacturers. I think this section

founder of the order his views would have seemed to be entitled to weight.

There is no evidence that his warning criticism of a narrow policy was heeded. The Grand Lodge and its officers from first to last seldom influenced the rank and file of the Order of St. Crispin after the initial constitution was once made and accepted. The Order's lack of flexibility in views, and lack of stability in action, were its own undoing.

Effects of St. Crispin Organization upon the Boot and Shoe Industry

Temporary success was all that the Crispins ever gained in Massachusetts. Burrell and Maguire, of Randolph, acceded to the first demands of the St. Crispins, and then closed its shop.[1] One Abington firm, Washington Reed's, went out of business rather than submit, it is said; but it sold out to another firm who accepted none of the demands and did a flourishing business. Various shoe men have told me that neither manufacturers nor non-members of the order took the organization, or its demands, very seriously. Another, an active officer of the Randolph order, told me in lively disgust, forty-five years after the Order closed its

should be materially changed, modified, or more clearly defined. In several places in the Western States where the manufacturing is confined to custom work, lodges have got into some trouble with the manufacturer, and their best and leading men have been obliged to leave town or open a shop of their own. The cases are quite numerous where members have opened a little shop of their own, have been obliged to leave the order as soon as so doing, which has been detrimental to the lodges and weakened it very materially. From the commencement of our organization, I have never been able to see the necessity of asking a member to withdraw from the Order because he happens to be fortunate enough to rise to foremanship, or even start a little business for himself. I contend it does not follow that a member cannot be a good member still under these circumstances as long as he is willing to be governed by our laws. When we get good men, I believe it is our duty to keep them if we can. We need more good men than we already have. I would go as far as to allow a custom manufacturer to join the Order. For, if they join, they must comply with our rules. If they do not join, there is nothing to prevent them from teaching new help as much as they please. It should be borne in mind that our western custom manufacturers are very different from those of the East, where it is more in the line of wholesale. They are all practical men or nearly so, and are fully qualified to teach new help, which is the very thing we are trying to prevent."

[1] See Appendix XXXVII.

career, that some fellow workmen were always cutting the price on the sly, and helping out the manufacturers either directly or indirectly. After the Chinese labor threat had once been carried out, the opposition to green hands became milder, for the manufacturers had both won the victory and created a fear. The shoemakers who stayed out of the Order have told me there was always work enough for all; that manufacturers hired in green hands or shut down entirely only when demands of the St. Crispins were too unreasonable to allow for profits. The Massachusetts shoemakers never had the rivalry of contract prison labor to compete with as their fellow Crispins did in other states, notably New York and Pennsylvania, and they never actually suffered as those of California did from the Chinese labor. The "green hand" opposition was less intense as the shoemakers grew older, realized they would have to be replaced and that probably all their sons would not want to learn to be shoemakers. This sentiment was voiced by Mr. S. P. Cummings, International Grand Scribe, in the meeting of the Fifth Grand Lodge gathered in Boston in 1872.

There is another point which I wish particularly to call your attention to, and that is the learning of new "Help." We may all consider that necessary to our welfare, but there are many circumstances which demand that this objection should not be tolerated, as for instance the learning of the orphan of him who died in his country's cause, as also the learning of the children of other Trade Unionists, who may fancy our craft. These considerations I earnestly recommend to you. . . .

No doubt but that there are many members of this convention who think, while we provide against the admission of new help, that in time we can control the price of labor, or, in other words, sell our labor for whatever we please. To those who hold such views I will refer them to the history of trades unions in Germany, France, and England, and you will see this curtailing the surplus help is nothing new with us. In fact, it is as old as trades unions themselves; it has always existed, and I trust it always shall exist, but in a different form than we have it at present in the Crispin Order. I am well aware there are those who are opposed to making any innovation upon what are considered the fundamental principles of our Order. However, I am satisfied in my own mind that the time is not far distant when we must adopt some law for apprentices; for I claim it will be more to our advantage to learn the youth of our country than to have foreign nations do so for us. Now, what benefit shall we derive if we should never learn another shoemaker in this country? To my mind this question seems plain enough — that we would gain but very little by such a course of action, for European markets would then become the school of instruction for the future shoemakers of America. Now, if all trades adopted our rule in regard to help, in another generation there would be no such thing as an American mechanic in the country. I suppose the question will be asked, If the prevention of new help does not maintain the price of our labor, what will? That, I think, is the question foremost in the minds of many of our members at the present time, and that is what I have often asked myself, and I have arrived at this conclusion: that in order to restore confidence to the minds of our members we shall have to convince them that it is for their interest to belong to such an order as this. Now, how shall we be able to convince them that they will be benefited by a union of this kind? I would say, make a reduction of wages a grievance, then every member will feel that he has something to lean upon in the hour of his adversity. I feel confident, myself, that, unless we shall be able to make some such grievance as this, we will cease as an international body altogether; and I would, in all sincerity, ask of the delegates present, aye, and every shoemaker in the country, What do you think would be the consequences then? Why, simply this: for a short period of time some of the lodges would have a local existence, and disappear like the dew drops on the hillside before the morning sun. Such, my brothers, I honestly believe will be our fate, unless we make provisions to support lodges when they are fighting our battles against a reduction of wages. A reduction of wages in one section eventually means a reduction of wages over the whole country; and still how easy it is for us to prevent all this if we only support each other against this cut-down system which is practiced upon us so often. We will then be able to fight the capitalist with his own weapon, namely, capital. I hold that without money we can never expect, nor need we hope, to be successful. In fact, without money we have about the same chances of being successful with a capitalist as an unarmed mob would have to defeat an armed and well disciplined army. I trust this grievance question will receive due consideration from every delegate present, for upon this depends the life or death of our Order.

Decline of the Order

Thus by one of their chief leaders, the Crispins were told that the article of faith, the fight against green hands, which made the most telling appeal in 1867–70 when the Order was recruiting members, had been outgrown and should be put aside. At that same meeting, the same speaker unconsciously pronounced the eulogy of the Order.

Passing to the concluding topics of my report, I must say that I have no fears for the future of this Order, because it is founded in necessity and justice, and cannot die until the one is removed and the other forgotten.

Finally, Brothers, what is to be our future? Are we going up or down? Shall we grow weaker or stronger? Five years have passed since seven men organized the first lodge. Since then it has become international as a union, and has a world-wide reputation; but after all its battles and victories, its struggles and sacrifices, what has it accomplished? Accomplished! It has more than lifted into respectability the craft it represents; it has forced forward the discussion of the labor question a quarter of a century in five years! I have traveled largely over the country the last three years, mingled with lawyers, politicians, clergymen and editors, many of them with world-wide reputations, but have never yet seen the hour when I was ashamed to say, "I am a Crispin." The Crispin Order has wielded for five years the most powerful weapon avarice, love of power and selfish greed ever met, and though sometimes the weapon may have been turned aside or blunted, still, in the number of its battles and its victories, the Crispin Order stands without peer on the Continent. We know labor has been better paid, better respected, become vastly more intelligent because of our existence; we know how today labor and its rights is the absorbing topic of public and private discussion — all this the work of five years.

The "Hard Times of 1873" when manufacturers had to retrench or give up their business all together, coupled with the higher cost of living, felt most severely by the working classes, probably did much to complete the decline of the St. Crispin Order through making its members thankful to have work on any terms the manufacturers would make. The human element in the boot and shoe industry, however, has not ceased to be a problem and a factor to be reckoned with, in any organization of the industry.

Close of the First Factory Phase in 1875

The year 1875 found the boot and shoemakers of Massachusetts unorganized as laborers, but used to complex machinery, which continued to revolutionize the shoe industry both as to uniformity and amount of product. That year saw the industry safely over the "Hard Times of 1873," recuperating as usual by renewed specialization. This time the specialization was in the Goodyear Welt shoes made by a machine which, with its attendant inventions, have made a characteristic group in the development of a second phase of the Factory Stage, besides closing the distinctly McKay era and the adoption of the Factory System.

The development of the United Shoe Machinery Company on the side of capital, and the Boot and Shoe Workers Union on the side of labor, are the most interesting as well as the most vital characteristic facts or elements of the second phase of the Factory Stage, lasting from 1875 to our own day.

APPENDICES

APPENDICES

I

Processes on Shoes in a Modern Factory

Of the 100 or more operations of a modern factory, more than 50 may be performed by machines. The number of operations, both hand and machine, varies with the process and product and the equipment of the factory. These operations are listed and briefly described in the following pages of a bulletin of the United States Department of Labor on Wages and Hours of Labor in the Boot and Shoe Industry. No attempt has been made to explain or paraphrase them. They are given to present in bold relief a picture of the complexity of modern shoe making in contrast with the simplicity of the early craft.

A shoe factory usually has the following departments: cutting, sole leather, fitting or stitching, lasting, bottoming, finishing, and packing.

In the cutting department are cut the several parts of the uppers, the lining, and the trimmings. These parts pass to the fitting or stitching department, where they are sewed together, forming the whole upper.

In the sole-leather department the soles are cut, and heels, counters, and boxes made. Frequently these parts are bought ready-made from factories making a specialty of such manufacture.

In the lasting department the upper, insole, counter, and box are assembled and fitted together on the shoe last. From this department the lasted shoe is sent to the bottoming department, where the welt (in welt shoes) is sewed on, the outer sole sewed on, the heel attached, and the heel and the edge of the sole trimmed to shape and finished.

In the finishing department the shoe is smoothed with a hot iron, scratches rubbed down, stains removed, and the shoe given a final cleaning and inspection. From this department the shoes go to the packing department, where they are boxed and cased for shipment.

The occupations for which data are shown are here listed in alphabetical rather than process order, under each department. The departments, however, are listed in process order.

Cutting department:
  Cutters, lining, cloth, male.
  Cutters, vamp and whole shoe, hand, male.
  Cutters, vamp and whole shoe, machine, male.
  Skivers, upper, machine, male.
  Skivers, upper, machine, female.
Sole-leather department:
  Channelers, insole and outsole, male.
  Cutters, outsole, male.
Fitting or stitching department:
  Backstay stitchers, female.
  Button fasteners, female.
  Buttonhole makers, female.
  Closers-on, female.
  Lining makers, female.
  Tip stitchers, female.
  Top stitchers or undertrimmers, female.
  Vampers, male.
  Vampers, female.
Lasting department:
  Assemblers, for pulling-over machine, male.
  Bed-machine operators, male.
  Hand-method lasting-machine operators, male.
  Pullers-over, hand, male.
  Pullers-over, machine, male.
Bottoming department:
  Buffers, male.
  Edge setters, male.
  Edge trimmers, male.
  Goodyear stitchers, male.
  Goodyear welters, male.
  Heel breasters, male.
  Heel burnishers, male.
  Heelers, male.
  Heel scourers, male.
  Heel-seat nailers, male.
  Heel sluggers, male.
  Heel trimmers or shavers, male.
  Levelers, male.
  McKay sewers, male.
  Rough rounders, male.
Finishing department:
  Treers or ironers, hand, male.
  Treers or ironers, hand, female.

Cutting Department

All operations of the cutting department here shown are usually performed by men, except skiving, on which operation women also are employed.

Cutters, Lining, Cloth. — Included in this occupation are the men who cut the cloth lining of the upper of the shoe. The work may be performed by hand or machine. The hand cutter receives the cloth folded 8 to 12 thick. He lays his patterns on the cloth and draws a knife along the edge of the pattern, cutting through the several thicknesses of cloth. The machine operator uses a die which, under the pressure of a power machine, cuts 24 to 32 thicknesses at one time.

Cutters, Vamp and Whole Shoe, Hand. — This occupation includes the men who cut by hand the entire top or outside of the shoe. It covers the men who cut the vamp and possibly some or all of the other part of the top also, but does not include cutters of minor parts only.

The vamp is the most important part of the upper and requires the greatest skill in cutting. It consists of the part or parts of the upper attached to the sole. The upper, according to the style of the shoe, may have other parts also, as quarters, tongue, tip, backstay, and foxing. The operator has a bench upon which he spreads the skin; he lays the pattern in the desired place and draws a knife along the edge of the pattern, cutting the part to the desired shape. For each different part of the upper there is a separate pattern. Incidental to the outside cutting care must be exercised in selecting like qualities and weights of stock for the same parts in a pair of shoes.

Cutters, Vamp and Whole Shoe, Machine. — These operators cut the same part or parts as the hand cutters described above. Instead of patterns and a knife they use dies operated by a power press. Different dies are required for each part of each style and size of shoe. The cutting board is similar to that used by the hand worker, with a beam over it which can be swung either to the left or right and any position over the board. The cutter places the die in the desired position on the leather, grasps the handle of the beam of the clicking machine and swings it over the die, with a downward pressure. A clutch is placed in operation, which brings the beam downward, pressing the die through the leather. After the cut the beam automatically returns to its full height and remains there until the handle is pressed again.

Skivers, Upper, Machine. — Skiving consists of cutting away, on the flesh side, the edge of a piece of leather, so that the edge may be turned and pasted back, thus giving a finished rather than a raw edge of the same thickness as the other parts of the leather. The machine used has a sharp-edged revolving disk so shaped as to cut the desired bevel or shoulder on the leather fed to it.

Sole-Leather Department

The two operations of this department for which data are shown are performed by men.

Channelers, Insole and Outsole. — The operator has a machine that cuts a slit near the edge of a welt insole or a McKay outsole. The slit extends only part way through the sole and is cut at an acute angle. The lip or lid of the channel is turned back by a channel turner. The channel in the outsole of the welt shoe is cut by the rough rounder. The channel in the welt insole avoids a seam inside the shoe and permits the insole, the welt, and the upper to be stitched together while on the last. In the outsole the channel permits the seam to be countersunk or embedded in the sole. After the outsole is stitched on, the lip of the channel is cemented down on the thread, protecting it from wear.

Cutters, Outsole. — The operator cuts the outsole from a side of leather by means of a die and a heavy descending power beam. The leather is laid upon the cutting table, the cutter places the die, and with his foot presses a lever, releasing the beam, which comes down upon the die with sufficient force to press it through the leather. The operator sets the die to have as little waste leather as possible, and to have the same quality of leather in a sole.

Fitting or Stitching Department

Women are employed so generally in this part of the manufacture of a shoe that data are shown for females only in all occupations reported except vamping, in which occupation wages are shown for both sexes. All of these are machine operations.

Backstay Stitchers. — The back of the shoe is usually strengthened by an additional strip running all or part of the way from the top to the sole of the shoe. This stay is sewed on over the back seam. This work is also called back stripping.

Button Fasteners. — This is an automatic machine operation. The buttons are fastened on the shoe by either thread or wire. The upper comes to the operator with the position marked for each button. The operator has only to put the top of the shoe in position and start and stop her machine.

Buttonhole Makers. — The upper is received by the operator with the position of each buttonhole marked. The machine cuts and works the buttonhole automatically. The operator has only to hold the upper in position and control the machine.

Closers-on. — This operation consists of stitching the lining to the top of the upper, both of these parts having been made previously. The work is also called inseaming. This operation is not performed on all shoes. In some shops it is omitted entirely; in others the lining is pasted on, which holds it to the top until the upper is top-stitched.

Lining Makers. — These employees are sewing-machine operators who sew together the several parts of the shoe lining. In some shops the work is subdivided, two or more persons doing a part of the work on each lining. This operation of sewing the cloth lining requires less skill than the leather-sewing operations, to which lining makers are usually advanced as they acquire skill.

Tip Stitchers. — The tip is a separate piece of leather generally put over the toe of the shoe. It is stitched to the vamp by machine.

Top Stitchers or Undertrimmers. — When the lining has been closed on to the top of the upper, it is folded inside of the upper covering the closing-on seam and passed to the top stitcher who stitches, by machine, the edge of the folded-in seam. This operation is also called undertrimming. In some shops this top stitching is done without a previous closing-on, the lining being held in position or previously pasted.

Vampers. — The vamp is the part of the upper to which the sole is attached. Vamping is the process of sewing together the lower part of the shoe, or vamp, and the upper part, known as top or quarter. Pumps or slippers having no tops or quarters do not require vamping. Vamping is the most important and best-paid operation in the fitting room. The operator uses either a single or double needle power vamp sewing machine. Either males or females may operate the machine; women's shoes being light can be vamped by females, but men's shoes being heavier require male vampers usually; however, in many plants making men's shoes only, there are women vampers.

Lasting Department

Men are employed almost exclusively in the operations of the lasting room.

Assemblers, for Pulling-Over Machine. — The assembler receives the last with the insole tacked on it. He wets the leather, shellacs the toe box or the tip or both, places the toe box and counter between the lining and the upper, and then puts the last inside the upper. Having centered the upper on the last, he places the last on the spindle of the assembling machine. By pressing a lever the machine automatically drives small tacks through the upper and insole into the bottom of the last at the toe, the heel, and either side, the tacks holding the upper in place temporarily. The shoe goes from the assembler to the machine puller-over. When the pulling-over is a hand operation, the assembling is done by the hand puller-over.

Bed-Machine Operators. — Lasting is the next operation after the shoe has been pulled over the last. The bed-machine operator places the shoe on the machine and by levers moves a series of wipers (friction pullers) which draw the upper over the edge of the insole at the toe and heel. Some factories designate this as toe and heel lasting. The shoe is placed with the sole up and the operator determines whether the shoe is properly lasted by placing his hand under the toe or heel. The wipers are kept in motion until the operator is satisfied that the upper has been wiped into the desired position. Under the welt system, the operator drives a tack through the upper and insole and partly into the last at one side and passes a fine wire from it around the drawn-in upper at the toe to the opposite side of the last and drives a tack, around which he winds the wire. The wire holds the toe of the upper in position as drawn in over the last. Under the McKay system, instead of the wire used on the toe of welt shoes, tacks are used. The upper at the heel is fastened by tacks driven in by hand. In case the side is lasted by the bed-machine laster the side or instep is lasted by hand with pincers. The operator draws the upper tightly over the last so that there are no wrinkles and tacks it down by hand. Lasting is one of the most important operations in the making of a shoe.

Hand-Method Lasting-Machine Operators. — In this method of lasting, which is done on a machine known either as "consolidated" or "niggerhead," the operator holds the edge of the shoe so that the pincers of the machine grasp the upper and draw it evenly and closely about the last. Immediately following the pincers as fast as the upper is drawn into position, there is a device on the machine that drives tacks automatically into the last to hold the upper in its proper place. In case any part of the shoe has not been properly lasted, the operator pulls the tacks and does the work over. Under the welt system this machine is often used to last only the side or instep while the bed machine lasts the toe and heel, thus cutting out the operation of hand lasting the side or instep, which is necessary in plants using only the bed machines.

Pullers-Over, Hand. — The hand puller-over is his own assembler, which occupation is described above. With the parts assembled, he takes hand pincers and draws the upper over the last and insole, taking care that the upper keeps its proper position, and drives a tack at the toe and two on either side to hold the upper in position for the laster.

Pullers-Over, Machine. — Where shoes are pulled over by machine, they are first assembled and put over the last by the assembler. The machine-puller places the shoe in the machine and the pincers of the machine grasp the leather at different points on each side of the shoe. The operator stands so that he can see when the upper is properly centered. He presses a foot lever closing the pincers, which draw the leather securely against the last. The machine stops at this point and the operator can start or stop the machine at will. The operator now examines the shoe to see whether all the parts have been evenly pulled over the last. Where a part has not been properly pulled over it can be adjusted to the desired point by levers. When satisfied that the shoe is properly adjusted, the operator presses a foot lever, the pincers move toward each other, drawing the leather around the last, and at the same time the machine automatically drives two tacks on each side and one at the top through the upper and insole into the last to hold the upper in position.

Bottoming Department

All operations in the bottoming department are usually performed by men.

Buffers. — After the shoe has been bottomed, the buffer removes stains from the sole and gives it a smooth, finished appearance by holding it against a revolving roll or wheel covered with sandpaper or emery paper.

Edge Setters. — The edge setter holds the edge of the sole against a machine having hot irons shaped to fit the edge of the sole, which irons vibrate rapidly and give a lasting polish to the edge.

Edge Trimmers. — The operator holds the edge of the sole against a machine having a series of revolving knives that trim the edge smooth and to the desired shape. This operation comes after the bottom has been sewed on and precedes edge setting.

Goodyear Stitchers. — The operator uses a Goodyear outsole lock-stitch machine to stitch the outsole to the welt. The seam is run in the channel in the outsole through both outsole and welt on the outside of the shoe. The stitches show on the upper surface of the welt and are covered on the under surface of the sole by cementing down the lip of the channel.

Goodyear Welters. — The welt is a narrow strip of leather to which the outsole is to be stitched. It extends around the edge of the shoe as far back as the breast of the heel. By one operation of the machine both the upper and the welt are sewed to the insole, the thread passing through the slit of the channel in the insoles. The outsole is stitched to the welt in a later operation.

Heel Breasters. — The heel breaster operates a machine having a knife which cuts to shape and trims evenly the breast or front surface of the heel, cutting down to the outsole, but not cutting into it.

Heel Burnishers. — The final operation on the heel is the burnishing. The operator holds the shoe with his hand in such position that the heel comes in contact with a wheel on the burnishing machine, which gives it a hard, smooth surface. Hot wax is carried to the heel by a small disk and applied by a series of rubbing blows, which beat the wax thoroughly into the heel. A revolving brush on the same machine brings the heel to a perfectly smooth surface.

Heelers. — The heels come to the heeler ready-made, except for the top lift or last layer of leather. A helper sticks nails in a steel plate. The heeler places the shoe on a jack or metal last, puts the heel in position, swings the nail plate into position over the heel when the nails are dropped into another plate over the heel. By operating a foot lever another part of the machine drives the nails down through the heel, the insole, and the upper folded between the insole and the heel, and clinches the nails back into the leather of the insole. The nails protrude slightly above the unfinished heel. The top lift, coated with cement, is then pressed down by the machine on the protruding nails.

Heel Scourers. — This operator holds the shoe by hand so the heel, trimmed but yet not smooth, comes in contact with rolls covered with sandpaper, which smooth the heel. The next operation on the heel is burnishing.

Heel-Seat Nailers. — The heel seat is the heel end of the sole. The insole, the outsole, and the part of the upper brought in between them are nailed together by machine. Small brass nails are driven automatically through the parts and clinched on the insole side. The shoe is placed on a jack and the work of the operator is to guide it during the nailing.

Heel Sluggers. — The slugger operates a machine which drives small pieces of brass or other metal, called "slugs," into the toplift of the heel to protect it from wear. The operator of this machine adjusts the plate so as to place accurately the desired number of slugs, and the machine automatically cuts off and drives the slugs as they are drawn from a coil of wire.

Heel Trimmers or Shavers. — The heel, when the shoe is received from the heeler, is rough and larger than the required size. This operator holds the shoe by hand in such position that the heel comes in contact with a series of revolving knives on his machine which cut away the heel to conform to the desired contour, as indicated by the top lift, which top lift is of exact size when put on by the heeler. The machine has two sets of knives. With the first set the trimmer shaves that part of the heel from the top lift to the sole, then with the other set he trims the edge of the sole, taking care not to cut the upper.

Levelers. — The operation of leveling to correct any unevenness in the bottom of the shoe is done with an automatic sole-leveling machine. The operator places the shoe on a jack or metal last, which he attaches to the machine, where it is securely held by the spindle and a toe rest. He presses a foot lever and the shoe passes automatically beneath a roll under heavy pressure. This roll moves with a vibrating motion over the middle of the sole of the shoe from the toe down to and into the shank and passes back again to the toe. The roll then cants to the right and repeats the operation on that side of the sole, returning to the toe as before. It then cants to the left, repeating the operation on that side, after which the shoe automatically drops forward and is relieved from the pressure. While one shoe is under pressure the operator is preparing another shoe for the operation.

McKay Sewers. — This operator uses a McKay sewing machine to sew together the outsole, the upper, and the insole — the three parts being sewed together in the one operation, except the heel seat, which is nailed. No welt is used in the McKay process, the seam being embedded in the channel of the outsole; the opposite side of the seam is on the inside of the shoe instead of on the top of the welt outside the shoe, as in the welt process of manufacture.

Rough Rounders. — This operation consists of trimming by machine the edge of the outsole and welt so that they will extend a uniform distance from the upper. It is the first operation on the edge of the sole in the Goodyear process. The machine also cuts a channel in the outsole, in which the thread is embedded, when the Goodyear stitching is done later.

Finishing Department

Treers or Ironers, Hand. — The treer places the shoe on a form, the shape of a last, supported on a frame. By pressing a foot lever the form is expanded so that the shoe fits tight over it. The tools of the treer are a hot iron, brush, cloth, etc. The treer brushes the shoe, cleans spots and discolorations, remedies any slight cut or blemish, and rubs the upper with a hot iron to take out wrinkles and produce a smooth surface. — *Bulletin of the United States Bureau of Labor Statistics,* No. 134, pp. 22-31.

II

Modern Shoe Repairing

[Not only in making but in repairing shoes in these days are elaborate machines deemed necessary so that the old-time cobbler cannot even repair a shoe with his old kit now in competition with such a repair outfit as this one described below.]

American ingenuity was exemplified this afternoon in the presentation by George W. Brown, vice-president of the United Shoe Machinery Company, to Brig.-Gen. Sweetser's command of a complete shoe repairing unit, mounted on wheels. The truck and trailer is the first equipment of its kind provided for the comfort of the infantryman in any of the armies of the world.

Mr. Brown presented the motor-driven vehicle to the brigade on the muster field, in the presence of Gen. Sweetser and his staff. He said, in part:

When the United States entered the war, our president, E. P. Brown, gave directions that the equipment that you see here be constructed, that the Massachusetts troops might have the benefit of it. Any effort along this line is essentially experimental. It is much more of a problem to repair shoes by machinery than it is to manufacture new shoes of whatever grade by machinery.

This equipment has a capacity of resoling approximately 400 pair of shoes daily. It has storage place for all the materials necessary for protecting its supply department. It is equipped with all the duplicate parts necessary to prevent a delay in operations incidental to wear, tear and breakdown. This applies to repairing machinery, electrical equipment, motor-truck, and trailer.

It is thought by military men that the greatest need of repairing machinery would be immediately after the unit to which it is attached had completed a march. In order that it may be immediately available at that time, sleeping accommodations have been provided, that the men who operate the machines may sleep while the army is marching, and be ready to operate the machinery during the entire night following a march. Means are provided for lighting the entire equipment by electricity.

Gen. Sweetser, in accepting the equipment, said, in part:

On the foot efficiency of the soldier depends the ultimate success of the army in the field. The seasoned soldier, once he has broken in a pair of shoes, can travel farther and better if they are kept in good condition, and I consider this outfit which the United Shoe Machinery Company has offered to be of inestimable value.

APPENDICES

As far as I can ascertain, this repairing outfit is the first of its kind in our country, if not in the world. It is really of greater value that the soldiers have such an apparatus, which will ensure them comfortable shoes, than to have an unlimited supply of new shoes available. From an economic viewpoint this shoe-repairing outfit serves an important function in the conservation of material. — *Boston Herald*, August 15, 1917.

III

MEDIAEVAL SHOEMAKING TOOLS

The bones of St. Hugh, a shoemaker who became a martyr, were found hanging in a tree by his comrades. Out of them, they devised, according to tradition, a full kit of tools necessary for a mediaeval shoemaker.

> My friends, I pray you listen to me,
> And mark what S. Hugh's Bones shall be.
>
> First, a Drawer and a Dresser,
>   Two Wedges, a more and a lesser;
> A pretty Block Three Inches high,
>   In fashion squared like a Die.
>
> Which shall be call'd by proper Name,
>   A Heel-Block, ah, the very same:
> A Hand-leather and a Thumb-leather likewise,
>   To pull out Shoe-thread we must devise.
>
> The Needle and the Thimble shall not be left alone,
>   The Pincers, the Pricking-awl and Rubbing Stone
> The Awl, Steel and Tacks, the sowing Hairs beside,
>   The Stirrop holding fast, while we sow the Cow-hide,
> The Whetstone, the Stopping-stick, and the Paring knife,
>   All this doth belong to a Journey-man's Life:
> Our Apron is the Shrine to wrap these bones in;
>   Thus shroud we S. Hugh's Bones in a gentle Lamb's Skin.

> Reprinted from
> *The Delightful Princely and Entertaining History of the Gentlecraft.* Published in London, 1725. pp. 55–56.

IV

Partial Contents of the Delightful Princely and Entertaining History of the Gentle Craft

- VIII. How the Lady Ursula finding herself to be with Child, made great Moan unto her Husband Crispine, &c.
- IX. How fair Ursula came before her Father with Crispine her Husband, who was joyfully received by him, &c.
- X. How Sir Simon Eyre being at first a Shooemaker, became in the End Lord-Mayor of London, thro' the Counsel of his Wife, &c.
- XI. How Simon Eyre was sent for to my Lord-Mayor's to Supper, and shewing the great Entertainment he and his Wife had there.
- XII. How John the Frenchman fell in Love with one of his Mistress's Maids.
- XIII. How Master Eyre was called upon to be Sheriff of London, and how he held his Place with Worship.
- XIV. How Haunce having circumvented John the Frenchman's Love, was by him and others finely deceived at the Garden.
- XV. How Master Alderman Eyre was chosen Lord-Mayor of London; and how he feasted the Apprentices on Shrove-Tuesday.
  A New Love Dialogue between a Young Lady and a Shooemaker.
  A Shooe-maker's Widow's Question to her Man whom she fell in Love with. His Answer.
- XVI. Of the Green King of St. Martins, and his merry Feats.
- XVII. How the Green King went a Walking with his wife, and got Anthony Now Now to play before them, in which sort he went with her to Bristol.
  The Shooe-maker's Glory; Being a Merry Song in the Praise of Shooe-maker's, to be sung by them every Year on the 25th of October.

V

A Contemporary Account of New England Trades in 1650[1]

All other trades have here fallen into their ranks and places to their great advantage; especially Coopers and Shoemakers who had either of them a Corporation granted, inriching themselves by their trades very much. . . . As for Tanners and Shoemakers, it being naturalized into these occupations, to have a higher reach in managing their manufactures than other men in New England are, having not changed their

---

[1] Cf. Edward Johnson: Wonder Working Providence of Sion's Savior in New England, pp. 207-209, Book 3, Chap. 6, in edition of 1654.

nature in this, between them both they have kept men to their standard hitherto, almost doubling the price of their commodities according to the rate they were sold for in England, and yet the plenty of Leather is beyond what they had there, counting the number of people, but the transportation of Boots and Shoes into forraign parts hath vented all however; as for Tailors, they have not come behind the former, their advantage being in the nurture of new fashions; . . . Carpenters, Joyners, Glaziers, Painters follow their trades only; Gun-smiths, Locksmiths, Blacksmiths, Naylers, Cutlers, have left the husbandmen to follow the Plough and the Cart, and they their trades, Weavers, Brewers, Bakers, Costermongers, Feltmakers, Braziers, Pewterers and Tinkers, Ropemakers, Masons and Tile makers, Cardmakers to work and not to play, Turners, Pump makers and Wheelers, Glovers. . . . and Furriers are orderly turned to their trade, besides divers sorts of Shopkeepers and some who have a mystery beyond others, as have the Vintners.

VI

THE CHARTER OF THE COMPANY OF SHOOMAKERS, BOSTON, 1648

Vppon the petition of the shoomakers of Boston & in consideration of the complaynts which haue bin made of the damag which the country sustaynes by occasion of bad ware made by some of that trade, for redresse hereof, is it ordred, & the Court doth hereby graunt libtie & powre vnto Richard Webb, James Euerill, Robt. Turner, Edmund Jackson & the rest of the shoomakers inhabiting & howskeepers in Boston, or the greatest number of them, vppon due notice giuen to the rest, to assemble & meete together in Boston, at such time and times as they shall appoynt, who beinge so assembled, they, or the greater number of them, shall haue powre to chuse a master, & two wardens, with fowre or six associates, a clarke, a sealer, a searcher, & a beadle, with such other officers as they shall find nessessarie; & these officers & ministers, as afforesd, every yeare or oftener, in case of death or departure out of this jurisdiction, or remoueall for default, which officers & ministers shall each of them take an oath sutable to their places before the Gounor or some of the magists, the same being pscribed or allowed by this Court; & the sd shoomakers being so assembled as before, or at any other meetinge or assembly tobe appoynted from time to time by the master & wardens, or master or wardens with two of the associats, shall haue power to make orders for the well gouerninge of theire company, in the mannaginge of their trade & all the affayres therevnto belonging & to change & reforme the same as occasion shall require & to añex reasonable pennalties for the breach of the same; provided, that none of theire sd orders, nor any alteration therein, shalbe of force before they shalbe pvsed & allowed of by the Court of that county, or by the Court of Assistants. And for the better executing such orders, the sd master & wardens, or any two of them with 4 or 6 associats, or any three of them, shall haue power to heare & determine all offences agaynst any of theire sd orders, & may inflict pennalties pscribed as aforesd, & assesse fines to the vallew of forty shillings or vnder for one offence, & the clarke shall giue warrent in writinge to the beadle to leuie the same, who shall haue power there vppon to leuie the same by distresse, as is vsed in other cases; & all the sd fines and forfeitures shall be imployed to the benefit of the sd company of shoomakers in generall, & to no other vse. And vppon the complaynt of the sd master & wardens, or their atturny or advocate, in the County Court, of any pson or psons who shall vse the art or trade of a shoomaker, or any pt thereof, not beinge approued of by the officere of ye sd shoomakers to be a sufficient workman, the sd Court shall haue power to send for such psons, & suppresse them; provided also, that the prioritie of theire graunt shall not giue them precedency of other companies that may be graunted; but that poynt to be determined by this Court when there shalbe occasion thereof; provided also, that no vnlawfull combination be made at any time by the sd company of shoomakers for inhancinge the prices of shooes, boots or wages, whereby either our owne people may suffer; provided also, that in cases of dificultie, the sd officers & associats doe not pceede to determine the cause but by the advice of the judges of that county; provided, that no shoomaker shall refuse to make shooes for any inhabitant, at reasonable rates, of theire owne leather, for the vse of themselves & families, only if they be required therevnto; provided, lastly, that if any pson shall find himselfe greiued by such excessiue fines or other illegall pceedings of the sd officers, he may complayne thereof at the next Court of that county, who may heare & determine the cause. This commission to continue & be of force for three years & no longer, vnles the Court shall see cause to continue the same.

The same comissiom, verbatim, with the same libtie & power for the same ends, vpon the like grounds is giuen vnto Thomas Venner, John Millum, Samuel Bidfeild, James Mattocks, Wm. Cutler, Bartholomew Barlow, & the rest of the coops of Boston & Charlestowne, for the pventing abuses in theire trade. To continue only for three yars, as the former, mutatis mutandis. — *Records of the Colony of Massachusetts Bay in New England*, vol. iii, p. 132.

VII

EXCERPTS FROM THE TIMOTHY WHITE PAPERS

In a letter of September 15, 1725, before Timothy White went to Nantucket, he asked his sister for stockings, saying "let them be home spun, stone gray worsted; but if you cannot find such, get me a pair or two of sale stockings."

June	24, 1740	this Timothy White paid to Jos. Daws's wife for weaving 20/
July	9, 1740	Paid Maxcy for leather 20s.
July	15, 1740	Paid Richard Coffin 20/ for leather.
July	20, 1740	Paid Zach. Hoit Dr. to a pair of shoes 8/ and 1 Bush. Corn 8/6
August 21		Let Zach Hoit have 10/ and 19 day a pair of Shoes at David Clark's 20/.
August	1740	Paid Richard Coffin (by the Nav[l] Officer) for the Sole Leather 12/
April	1744	Paid to David Gardner for currying leather £0 5 4
May	25, 1744	Paid to Maxcy for Linings 15/
June	18, 1744	Paid Ruth Cromwell for wool-cash £9 0 0
May	1748	Paid David Gardner for a side of leather 70s

The Timothy White papers [1] show accounts with all the natives of Nantucket for the schooling of their children. Generally he was paid in money, but occasionally in cheese, corn, paper, fish, books, and rye. The fact of his taking cheese and corn in return for service and paying bills for the pasturing of his cow for many years suggests that he did not have much land to farm. Evidently school teaching was his main occupation, and making shoes for himself and family and neighbors a helpful resource.

VIII

EXCERPTS FROM THE GEORGE REED PAPERS

A picture of daily family life and household economy in this Home Stage of the boot and shoe industry is given, for example, in an ac-

---

[1] Included in the Timothy White papers reprinted by the Nantucket Historical Society.

count book of George Reed, of Dighton, Massachusetts. He wrote it from 1729 to 1763. Dighton was at least half a century behind the communities lying near Boston and the Bay in her economic and industrial development, so that the Home Stage of industry, when the community was almost self-sufficing and knew few markets, lasted until the third quarter of the eighteenth century. The book itself, of unruled, unglazed paper, with its untooled crinkled rawhide cover is suggestive of the home-made standards of this little inland community. The intimate, personal tone of the entries make for realism and for sympathetic attention.

Dighton march the 21 day year 1729. Thomas Reed

	£	s.	d.
and for one days work and tow shillings more	3	4	0
and for money lent	0	2	0
and for iron work	0	10	0
and for vamps for a pare of shoes and soles for yourself	0	3	1
and for 3 barels of cydear	4	10	0

George Reed.

Though a farmer, Reed evidently did not own oxen, for he repeatedly hired them. He owned a horse which he "hired out" along with his three sons, and he made money from his orchard and pasture.

Dighton, Oct. the 17 day year 1729 Cornelos Jones.

	£	s.	d.
to me for apples which you had in the loer [1] orchit	0	8	0
and for six bushels of winter appels	0	6	0
and for my mares going to Tanton [2]	0	2	0
and for my mares going to Tanton	0	2	0
more for my mare to Stephens Mill	0	1	0
and for my mare threshing	0	2	0
and for my mares going to Tanton to carrey your wife	0	3	0
and for your cow pastorin 2 weeks	0	2	0
and for my mare going to mill tow times	0	0	9
Dighton, novembear Credar for goods which I had of you at one time	1	17	8
at another time, cap, gloves and paper	0	15	3
and for two gallens and three quarts of rum	1	2	0
and for 3 pounds and a half of sugar	0	4	4
and for an almeneck which I had	0	0	6
and for a peck of pegs	0	3	9

[1] The phonetic spelling of the country dialect is warranted to give atmosphere to these naïve accounts even in reprint.

[2] Taunton is directly north of Dighton, and was then as now the nearest town. It was probably the chief market for the Dighton farmers. See map on page 14.

In the account for 1730 Reed appears not only as a farmer, but also as a quiet wit.

Dighton, June 13 day year 1730 Cornelus Jones debtor to me.	£	s.	d.
for killing a calf and for my mares going to Stephens mill one time	0	1	0
for my mares going to Tanton to carry your dear self............	0	2	0
for my mares going to Abbenton to carry your dear self..........	0	6	0
for my mares going to Willbor Daniel to by bef, sir, if you please..	0	2	0
for your pigs eating my corn, year 1730 you agreed..............	0	8	0
for my feching and caring granne hataway......................	0	2	0

For his son's work he got the following pay:

July 6 day year 1742. Daniel Whitman debtor.	£	s.	d.
to George helping cut stoks one day which was forgot............	0	2	0
to George going to mill.......................................	0	17	0
to half a days work about the Conor...........................	0	4	0
to one day of Johns Carting dung..............................	0	8	0
to Johns helping you ½ day about hey..........................	0	4	0
to George helping you on crib corn............................	0	3	0
to pastorin your oxen...	0	2	0
to Johnes helping you 21 days................................	4	0	0

At the same time, Daniel Whitman was keeping track of his indebtedness to this George Reed.

	£	s.	d.
Credate to your oxen to draw my hay to gather.................	0	3	0
to carting one day with one yoke of oxen......................	0	16	0
to your oxen one day...	0	10	0
to your helping me plow......................................	0	8	0
to your drawing one lode of wood.............................	0	6	0

Within the household, Mrs. Reed evidently gave some help in bringing in ready money, for in 1740, George Reed was charging to Dabora Talbot

	£.	s	d.
to eggs and thread and worsted and milk	0	9	8
to coming 4 pounds of wostord [1]	0	12	0
to coming 2½ pounds of wostord	0	6	0
to weaving 2 yards of cloath	0	4	1
to coming 2½ of wostord......................................			
for half a bushel of (?) at £1 15s. 0d. a bushel..................	0	17	6
for 9½ qts. of milk	0	9	6

In the Reed's kitchen, wool combing and weaving was done for the family and occasionally a neighbor, and the winter's supply of boots and shoes was evidently made up by the father and sons, for aside from

---

[1] Combing worsted.

the record of selling pegs and vamps there is no mention of buying either shoes or materials for them in all those pages covering the years 1729 to 1763.[1] All the other items are those which naturally accompany living in the Home Stage.[2]

IX

Some Typical Details of the Early Political, Religious, and Financial History of Four Massachusetts Shoe Centres

Randolph (Cochato)

Before 1715, there were settlers in the Cochato River valley of Old Braintree provided with log cabins and at least one saw mill, and by March, 1727, the South Precinct was set off from Braintree. Their petition to the General Court for such a distinct incorporation, dated December 28, 1727,

Humbly Sheweth That your Petitioners, and others of our Neighbors who joyn with us, are laboring under difficult, and distressing circumstances, in regard to the remoteness of our Habitations from the Publick worship of God for several of us dwell at Such a distance therefrom, that we, with our families are forced to travaillè upon the Sabbath five miles, Some Six, Some Seven miles to a Meeting to hear the word preached, upon such consideration, That it might be less labour and more easie for us, we have been at a charge to Erect a Convenient House, and have set it in such a Suitable place (tho. not, yet finished) as may very well accomodate the Neighborhood for Such a service. Also, we have chosen persons to seek a Suitable Minister to preach with us this Winter, this was done with the Advice and consent of our present minister Mr. Niles who has promised his assistance in this good work.

The General Court of Massachusetts and the town of Braintree granted this request to the pioneers of Cochato, and they were free to take up their social and religious independence. The political[3] dependence was acknowledged in the terms "South Precinct of Braintree" and the name Cochato was dropped in all official records.

[1] This account book is owned by Miss Julia Gilmore of North Raynham, Mass.
[2] Bücher, etc...., p. 89 of Wickett's trans.
[3] Precinct and town are distinct in their nature, object and powers. While the former is established solely for the purpose of maintaining public worship, its powers being limited to that effect, the latter, a civil and political corporation, is established for municipal purposes.

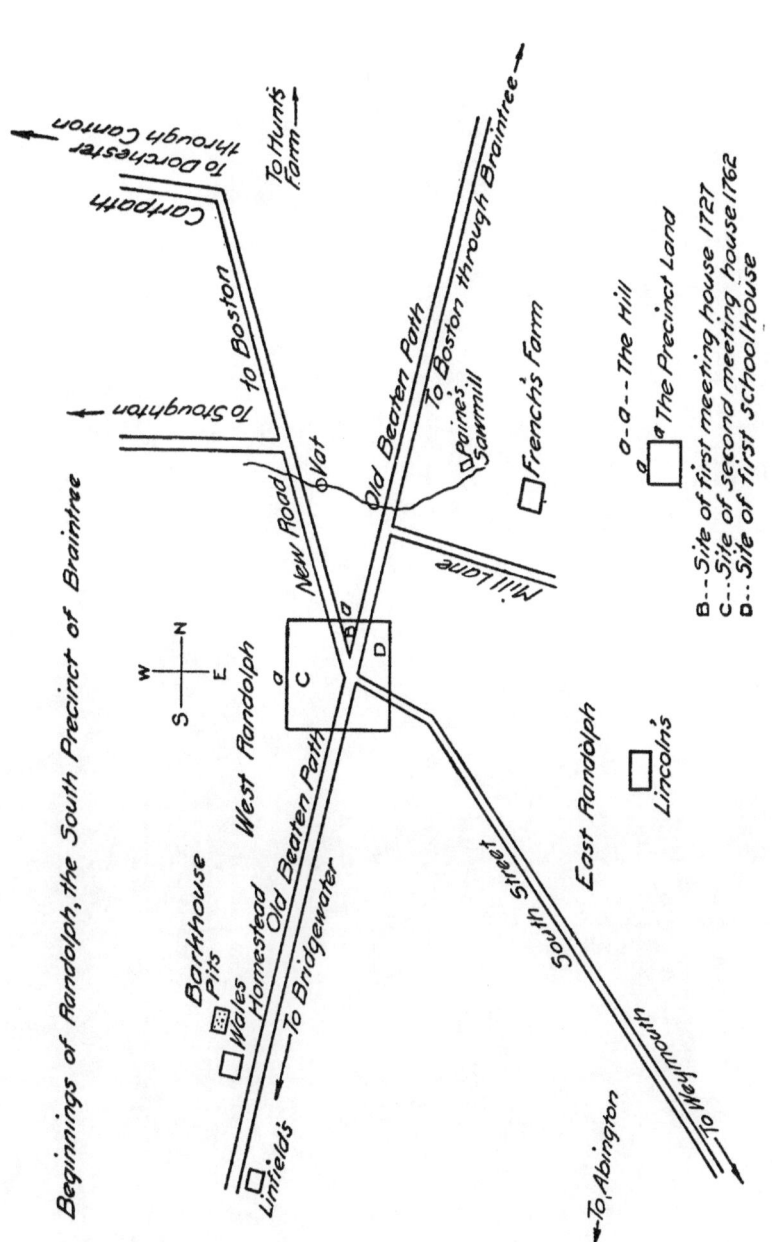

For this village and parish of "forty families and over 200 souls," a settled minister was secured with a "gift of £150 in money for settlement" and "£70 a year in money or Bills of Credit as Silver money at eighteen shillings an ounce." The young parish did not hesitate to assume large financial responsibilities, nor the new minister, Mr. Eaton, to accept a hard parish. From the meeting house on the "Hill" in the Precinct [1] Land, there led paths in various directions to the log cabins as scattered as the springs by which the settlers knew they must locate their farms. No fear of Indians influenced their action, or called for a village huddled about a central stockade. The accompanying map shows how scattered these farms were and makes one realize the difficulties of a village parson and the village doctor. "To many a cabin he must find his way by uncertain footpaths and trees marked by the axe." [2] From the Hunt farm to the French homestead was a distance of at least two miles in an air line.

The bridle path over which Samuel Sewall and Colonel Hawthorne traveled in 1704 [3] on the way from Boston to Taunton, "baiting" at Braintree and at Bridgewater had been given a formal location by 1714, and though it was still known as the "Old Beaten Path," its conventional name was hereafter the Bridgewater Road. From Braintree records, we can see how eagerly the young parish clamored for more roads to be built. It was not merely for the sake of easier traveling for the parson or any stranger's carriage,[4] but for the convenience of carts laden with farm produce, and lumber drawn by oxen, that roads were wanted by this young farming community.[5]

[1] "The central hill where the church and the schoolhouse were located, was known as the 'Precinct Land.'" Records show that Joseph Crosby, who lived in the North Precinct of Braintree gave to the South precinct for a consideration of forty shillings, a certain parcel of land in Braintree, in the County of Suffolk, in the Province of Massachusetts Bay, containing one acre more or less, situated on a country road leading to Bridgewater and set off by certain heaps of stones at the four corners thereof. This deed was dated March 1, 1727-28, and is entered with Suffolk Records, August 23, 1757, Lib. 90, p. 200.

[2] Cf. Discourse of Reverend John C. Labaree in Proceedings of the One Hundred Fiftieth Anniversary of the 1st Congregational Church of Randolph.

[3] Cf. The Diary of Samuel Sewall, vol. ii, p. 115.

[4] "It is related that for thirty years after the first meeting house was built, it was not approached by carriage." Cf. Benjamin Dickerman's Discourse in Proceedings, p. 65.

[5] Here is where the English colonists in America felt the lack of the old feudal organization which had furnished roads for England, or the later leadership of intelligent landlords with the means to suggest the way and bear the financial

After 1721 they sometimes took the cartway to Dorchester, going through what is now known as Canton. South Street, leading east from the Hill to Wessagusset (Weymouth) and to Abington, was located as early as 1731. Saddleback traffic might do for surplus supplies of yarn, linen, and woolen cloth but it would not do for lumber and vegetables. There must have been trade in surplus products, for otherwise, the product of the work of the men on the farms, and of women in the farm houses could not have been supplemented. How else could the parishioners raise the salary of the minister, and pay for the furnishings of the meeting house? No mortgage was ever put on any of the meeting houses set on the Hill, and no minister lost any of his salary. Then, besides supporting the Church, the South Precinct of Braintree had to pay its share of its mother town's public rates. It is by reading thus between the lines of the civil and parish records of the time that we get the economic history of the early New England towns. It is here also that we get our glimpse of the social and industrial life in the South Precinct of Braintree. Nothing is said of the clothes, warm coats, and rude shoes made in the kitchens at Cochato by the farmers' wives and the farmers themselves, but tradition supplies these details and adds that even in cold weather, the shoes were carried and not worn to church. It was only at a few rods distance from the Church door that the shoes were put on.[1] Not by all the Church-goers even then, for disturbance from the "barefoot boy" in the gallery was no uncommon feature of the Sabbath service. Such stories come to us by tradition along with others about men wearing frocks of blue homespun to meeting.

brunt in the rural community. This little settlement could only look to the state, *i.e.*, to political organization to supply them with economic necessities of a communal character. Yet the town, when its ear was once secured and its mind convinced, did its part. Cf. T. N. Carver, in the Yearbook of the Department of Agriculture for 1914, on the Organization of Rural Interests, pp. 240–400.

[1] Mr. Dow Barnes of Frenchtown, Pa., remembers distinctly how his Aunt Harriet came to their house each Sunday morning on her way from " Grandfather's Hill Farmhouse " to the meeting house, and stopped at his home to put on the shoes which she had carried so far in her hands. This memory concerns the years 1858 to 1863 in an isolated community where all shoes worn were made to order and treated with the same sparing care that the people of Randolph were giving to theirs a whole century and a half earlier. This seems like an interesting bit of evidence that economic conditions bring certain habits of thrift or the reverse. What man among us conscientiously or consciously saved his three dollar and a half machine-made shoes made before the World War even if he did tell his five-year-old son to look out for the toes of his?

The written records of the parish tell us details about the first, and then the second, meeting house. From the crudeness of this "best house" we may judge of the houses of the parishoners. Built two stories high, 44×32 feet, the meeting house was a plank building with a pitch roof, shingled and clapboarded, and had a double door for entering at the south. The broad aisle led from this door to the pulpit and seats to the right were occupied by men, and those to the left by women. Although the walls were lathed and plastered, there was not a bit of paint put on the inside or outside. The lumber had been sawed at Paine's Mill, down by Glover's Brook (see map), and had been put together by men of the parish according to the custom of all these early towns.[1]

If people then, in this frontier town, had to make their own share of the meeting house and their pews, it is but a natural part of the same economy that *they made their own boots and shoes*. Even when a man worked outside of his own farm for wages, the amount of ready money was small. Ten shillings was the pay for janitor service at the Church for the whole year of 1735. Twenty years later Seth Turner was paid only six pistareens[2] for a year's care of the meeting house, on the condition that he should sweep it twelve times, yet these Spanish coins were probably a large part of the ready money he had that year.

About the middle of the eighteenth century, the houses at the South Precinct were made more comfortable and looked less crude. Then some people thought that the community should see to it that the Lord's house should be made better. It took some years of discussion to make all the people ready for the new meeting house, or the cost it would entail. Perhaps the refusal to repair the old meeting house was the diplomatic move of a few leaders to procure an entirely new one. And so rain and snow found their way regularly into the church and yet the parish failed to vote for lime to join the gable into the meeting house and get two bundles of shingles and one hundred clapboards for repairs, in 1758.[3]

New settlers and increased wealth had evidently come to the village

[1] See town records of South Dennis, Dudley, Nantucket, North Bridgewater and Brookfield for similar proceedings and conditions. One vote of the Randolph parish reads that "John French is to build the floor in the gallery and the stairs leading to it, to lathe and plaster said gallery up to the beams, and for this work he is to have a pew in the gallery 7 feet long and 4 feet 10 inches in depth."

[2] In the Breed Papers quoted in Appendix XI, in the Daniel Parrott account, it might appear that a pistareen was equivalent to nine shillings in 1772.

[3] Discussion by Benjamin Dickerman in Proceedings . . . Randolph, p. 65.

by 1762, for, on January 14th, when all the freeholders and other inhabitants of the South Precinct were warned to meet at their meeting house at ten o'clock in the forenoon, it was there voted to build a new meeting house. The records of this new church show the fuller purses and the higher standards of the parish. Two hundred pounds were to be raised according to the vote towards building this meeting house and it was to be underpinned on the front and ends with two tiers of cut stone. The parish also voted to build a steeple fifty feet high to the bell deck and forty-six feet above that to the tip of the spire. The sills were of white oak, and cedar shingles, and cedar clapboards only were to be used. A sounding board was to be placed over the pulpit, and wide galleries were built on three sides. Evidently there was a chance for greater income from the Church-goers for the average price for a pew was about fifteen pounds, although they varied according to desirable location, from six to twenty pounds. The amount of ready money in the community can be judged by these figures. The cost of the new meeting house material amounted to little over £987. The money raised from the sale of pews, of the old meeting house and old school house, and that raised by the precinct and by church members from the neighboring town of Stoughton, had provided a total fund of £1012/8, so that the new meeting house started out unencumbered by debt and there was a balance in the treasury of £25/1/10. It was voted to sweep the meeting house oftener, to ring the curfew at nine o'clock each evening, and to toll the bell for the death of any parishioner or any member of the Precinct. Thus Randolph had come to division of labor and had already passed from the Home Stage of production and payment in kind into the Handicraft Period. In some industries, she was entering the Domestic Period, which would give more ready money to producers and lead to even greater division of labor.

Brockton (*North Bridgewater*)

The details of North Bridgewater's early history also include facts from the parish records which give an insight into their economic conditions in turn. By 1737, this village was ready to build a meeting house, to support a minister, and to control its own parish affairs. In the town meeting of 1739, Timothy Keith and Benjamin Edson were chosen a committee for the finishing of the meeting house in said precinct. It chose Abiel Packard "Recever of Stuf and meterrels." [1]

---

[1] Documents quoted in Bradford Kingman's History of North Bridgewater, pp. 20-25, 85-95.

By further agreement, James Packard was to do the glazing of the house, and "what it amounts to more than his Reats (*i.e.*, taxes) Come to, to take his pay at the forge, in Iron Ore or Cole next fall." John Johnson and John Kingman were to do the mason work, and provide the nails.

This North Bridgewater parish paid a bonus of £300 for the settlement of the minister, and agreed to pay the Reverend John Porter one hundred eighty pounds yearly in any passable money; his salary was to rise and fall as the price of silver did during the time that he should be the minister. By 1762, the parish had raised money enough for a new meeting house which had a belfry and a bell. Twenty-six years later, a parish committee reported as necessary repairs the following: "new sett the Glass in putty, paint the Door, windows, and walls, and the platform of the Belfree to be covered with led." This is our best evidences that the parish was growing richer. They had a comparatively rich and varied soil to depend upon as farmers. There were several brooks and a pond fed by a stream sufficiently deep to provide water power for their saw mills and grist [1] mills.

Brookfield (Quabaug)

. Stretching out over the Massachusetts lands from Boston towards the south were the settlements made at Braintree (1625),[2] Bridgewater (1639), Taunton (1637), and Plymouth (1620); towards the west were Concord (1635), Lancaster (1643), Sudbury (1635) and Springfield (1636). Between these groups of settlements there was a wilderness marked by more or less well defined Indian trails, and two English bridle paths which followed the divides and crossed the streams at natural fordways. Even before the Reverend Thomas Hooker started with his Cambridge (Newtown) congregation for the new lands to the west, in 1636, there was a definite trail towards the Connecticut River Valley known as the old Connecticut Path.

Brookfield as a frontier settlement had to repeat pioneer conditions. This meant for its settlers a return to the Home Period of production and hard economic struggles besides those with the Indians.

[1] Several men over fifty years of age living in Brockton have told me of going up Elm Street with corn to be ground at the old grist mill in their boyhood, and of their duty of watching the toll being taken out in kind. Now several factories stand on the grist mill site and the water is colored and scented with shoe paste.

[2] Cf. Records of the Massachusetts General Court; and Temple: History of North Brookfield for reprints of interesting source material and bibliography.

By 1660 white men from Ipswich were ready to take from the Indians this spot where the trails crossed, even if it were in the wilderness, provided they could get the consent of the General Court of Massachusetts. That was given, conditionally, in May of the same year in the terms of "six square miles near Quabaug Ponds" on condition of twenty families being resident there within three years and an able minister settled there within the said town. Indian troubles seem to have delayed such settlement until 1665, when neighboring grants gave rise to a fear of loss of claims. In desperation a few men put up houses and an actual settlement was begun. Needing a title from Indian owners, these settlers asked Lieutenant Cooper, of Springfield, to negotiate it. They collected three hundred fathoms of wampum, equal to seventy-five pounds in money, from the six or seven families there, and tried to get a new grant from the General Court of Massachusetts in 1667. Their remissness was reviewed, rebuked and forgiven by that body, and a regrant was made on similar conditions. The town had a hard time of it even after this, for all through the Indian Wars they were in trouble, loss, and terror from fire and tumult until at last, the whole village was wiped out in 1676. It was born again in 1686 and had its youth and early manhood before 1783.[1]

Lynn (Saugus)

Lynn was the oldest of the four shoe centres studied here in detail. Its early economic history can be gleaned, like that of the other three, only from stray sentences in the records concerned primarily with its political and parish life.

Founded in 1629 under the name of Saugus, by Edmund Ingalls, a farmer from Lincolnshire, England, it was intended for a farming community. The grants from the General Court of Massachusetts were generous in good lands near the Bay. Yet from the very first, the sea offered a rival to farming as an industry, in the way of fishing, and Francis Ingalls, brother of the founder, offered still another rival as early as 1630. Having been a tanner in England, he at once built a tannery at Humpfrey's Brook, in the part of Old Lynn now known as Swampscott. There the vats remained until 1823, to tell this fact if by chance the old men had forgotten the tradition of the first tannery in New England.

[1] These dates are for the first settlements and not for the incorporation of the towns.

The list of original settlers and landholders includes forty names.[1] Three were spoken of as captains, and another as lieutenant. They are recorded as having, as a community, herds and flocks of horned cattle, sheep and swine. From them they got not only their meat, tallow, butter and cheese, but wool and hides as well. From the earliest days they raised quantities of flax and rotted it in a pond which still has the name of Flax Pond.

With the coming of a clergyman, the Reverend Mr. Whiting of King's Lynn, England, there came a change of name for the Saugus settlement, the settlers taking the new name either to please him or to make him feel at home. When Edward Johnson's Wonder Working Providence was published in 1654, Lynn people found their town had been definitely described by him:

. . . her scituation is neere to a River, whose strong freshet at breaking up of Winter, filleth all her Bankes, and with a furious Torrent ventes itselfe into the Sea. This Towne is furnished with Mineralls of divers kinds, especially Iron and Lead. The forme of it is almost square, onely it takes too large a run in Landward (as most townes do.) It is filled with about one Hundred Houses for dwelling. Here is also an Iron Mill in constant use, . . . Their meeting-house being on a levell Land undefended from the cold North west-wind; . . . their streetes are straite and comly, Yet but thin of Houses. The people mostly inclining to Husbandry, have built many Farmes Remote there, cattell exceedingly multiplied. . . . Horse, Kine and Sheep are most in request with them." Bk. I, ch. 22.

X

Excerpts from the Southworth Papers

Even as late as 1811, conditions like those of Revolutionary days survived in shoemaking as a by-industry. Jedediah Southworth lived in West Stoughton about a mile from Stoughton Centre, on land that was set off originally to Capt. Consider Atherton.[2] He was a farmer who had cattle and horses to rent, and made as well as repaired shoes as a side occupation, using hides taken from his own cattle. His sons Luther and Apollos were "let out" along with his cattle.

---

[1] This, and many other interesting details of Lynn's early history, can be found in the official records of Lynn, and of the Massachusetts General Court. Cf. Lewis and Newhall: History of Lynn, for reprints of original sources and for bibliography.

[2] His great grandson, Loren Puffer, of Brockton, has the Southworth account books in his possession.

APPENDICES

Lewis Johnson[1] to J. Southworth, Dr.

1811

May 16	to Plowen one Day with three Cattle	$2.00
18	to Plowen one Day " "	2.00
21	to " " " " "	2.00
June 21	to Apollos one Day Hoeing	0.67
July 6	to Carten a L^d of Vinegar to Boston	4.00
16	to Carten five Barrel of Vinegar to B.	2.36
Aug. 3	to Carten a L^d of Hay from F. M.[2]	2.00
5	" " " " " "	2.00
7	" " " " " "	2.00
17	" " L^d of Vinegar to Boston	4.00

Lewis Johnson to J. Southworth, Dr.

1812

May 9	to Carten to Bos^n 5 bbls. of Vinegar, 5 doz. of Shovels one Set of wheels	$3.60
	to Carten up 4 Bushels of Corn and meat	.33
	to Sundry articles Brought up	.17
June 13	to Carten 4 lb. of Vinegar, 12 Doz of Shovels to Boston	3.60
	to Carten up Ten Bushels of Grain	0.53
	to " 4 barrels	0.33
	to " Herring flour molas and Sundry other articles	.40

L^t. Hez^h. Byay to J. Southworth, Dr.

1812

Feby 11	to drawen Seven Load of wood to W. Richmond	$5.33
Apr.	to drawen four Load of rails from E. Swamp	3.00

1813

Feby.	to four Days & half work Luther & Team	6.75
	to horse to Easton meeting	0.25
Mar. 10	to Carten up 15 1/3 of Salt Hay	2.00
	to 15 1/3 of Salt hay hundreds	7.87
Nov. 9	to three hhds of Lime	7.00
	to horses to Mr. Holmes	.20
22	to 37 cwt. of Salt Hay & Carten at fifty cents	16.50
	to Paying for weighing	.55
	to one Dollar on Moses Byay	1.00
		50.45

1814 Mar. 11 Settled

---

[1] Johnson — a farmer who made shovels.

[2] F. M. — Fowl Meadow, between Canton and Milton, right opposite Ponkapoag Pond. (See Randolph Map, p. 176.)

Contra to Hez.ʰ Byay

1812

	Recd 14 1/1 feet of Walnut wood	$4.75
	Recd 7 feet of Chestnut Bark	1.75
Nov. 6	Recd 6 feet of wood Standing on yᵉ Plain	1.50
22	Recd 6 feet of wood on yᵉ T. Pike[1]	1.88
27	Recd 6 feet wood on yᵉ Plain	1.88
	(etc. amounting to 21.25)	21.25

1814

Mar. 13	To Sawen & Slabs & S.	33.01
	Settled by balance	17.44
		50.45

James Smith, Jr. to J. Southworth, Dr.

1814

	Taken from Page 4	$4.75
Aug.ᵗ 27	to Jacket for Benj Clap	1.75
27	to two Jacket Paterns	3.00
Sept. 13	to Tapen Nathans Shoes	0.50
Oct. 30	to Eleven hhᵈ of Salt Hay	8.05
Dec. 13	to Tapen Nathuns Shoes and mending	.67

1815

Jan. 6	to a Pair of Shoes for Isaac Smith	1.83
18	to Tapen and Heel Tapen Nathan Shoes	0.75
25	to a Pair of Shoes for Ben Clap	2.00
Mar. 29	to Tapen and Heel tapen Benj. Clap's shoes	0.75
Sept. 26	to mending Nathan Dickeman's shoes	0.25
Oct. 23	to Carten up 7¼ of Salt Hay	1.45

XI

Excerpts from the Breed Papers

An account with Jeremiah Gray in 1768–69 would suggest that he was a journeyman whom Amos Breed sometimes paid by orders on other stores: —

June Ye 9, 1768

Jeremiah Gray to Amos Breed Dr.

		£	s.	d.
June 9	To 2 cakes of Sope	0	9	0
26	To a order at Zepaniah Breeds	1	1	0
	To 1 yard of cloath	1	4	0
July 9	To 1 pr. of shoes for self	2	15	0
	To a order at Zephaniah Breeds	0	8	0
1769				
Mar. 29	To 3 copers worth of thread		1	3

[1] T. Pike — Taunton Turnpike.

Another account, mainly about food and sole leather, was Daniel Parrott's. He was an independent shoemaker, evidently, from the supplies which he bought, and not a journeyman like the previous Gray. The account began in 1771.

		£	s.	d.
Nov. 11	To 9 pound of Beaff	0	14	3
	" prunes and Shougar at Zephaniahs	1	1	6
	" 1 pr. of vamps	0	2	6
23	" 1 pr. of vamps	0	16	0
	" Nife	0	3	0
1772				
Jan. 6	" cash paid, 1 pistoreen	0	9	0
	" sole leather	0	9	0
	" Prunes at Zephaniahs	0	6	0
Feb. 28	" cash paid for meal	0	17	0
	" cash paid for fish	0	9	0
Sept. 10, 1772.	To cash paid in full	55	2	3
	(Signed) Daniel Parrott.			

For the volume of Amos Breed's business at its greatest extent, we have such typical accounts as those of Ramsdell, Estes and Chase.

CR. TO MARSHACK RAMSDEL.

			£	s.	d.
1768					
Feb. 7	By 2 pr.[1] of Callam[2] Shoes		1	6	0
	" 5 " " " "		3	5	0
20	" 8 " " " "		5	4	0
27	" 6 " " " "		3	8	0
Mar. 5	" 7 " " " "		4	11	0
12	" 5 " " " "		3	5	0
19	" 6 " " " "		3	18	0
25	" 6 " " " "		3	18	0
Apr. 10	" 13 " " " "		8	9	0
May 14	Marshack went home to Bord[3]				
20	By 5 pr. of stuff shoes @ 12/		3	0	0
	By 5 pr. of stuff shoes		3	0	0
	" 7 " " " "		4	4	0
	" 4 " " " "		2	8	0

[1] This price was probably not for the shoes themselves, as the entry would indicate, but just for making the shoes, or else the shoe stock and work are both included and a low wholesale price much below that given for Breed's custom work.

[2] Calimanco, *i. e.*, calico.

[3] This would suggest that Ramsdell was a journeyman, boarding at Breed's until May 14, 1768. That would give weight to the belief that the pay was for work only and not for stock.

Cr. to Wm. Estes

1771			£	s.	d.
Nov. 11	By 6 pr. of stuff shoes @ 11/3		3	7	5
16	" 3 " " " "		1	13	9
23	" 5 " " " "		2	16	3
30	" 3 " " " "		1	13	9
Dec. 7	" 3 " " " "		1	13	9
14	" 3 " " " "		1	13	9
21	" 5 " " " "		2	16	3
28	" 3 " " " "		1	13	9

The following year, Benjamin Church was making more pairs at a time and bringing them in, as Estes had, about once a week.

1772		£	s.	d.
Oct. 31	By 5 pr. of Stuff shoes @ 11/	2	15	0
Nov. 7	" 4 " " " "	0	0	0
Nov. 21	By 2 pr. of Stuff Shoes	0	0	0
28	" 7 " " " "	0	0	0
Dec. 5	" 5 " " " "			
12	" 6 " " " "	0	0	0
19	" 8 " " " "	0	0	0
26	" 9 " " " "	0	0	0
June 2	" 7 " " " "	0	0	0
1773				
7	" 4 " " " "	0	0	0
16	" 4 " " " "	0	0	0
23	" 7 " " " "	0	0	0
These amount to		77	0	0

The contra account of Chase shows him buying shoemaker's tools and paying board.

Benjamin Chase to Amos Breed, Dr.

1772		£	s.	d.
Oct. 26	To 1 pr. of pinchers and Nife	0	11	3
	" 1 doz. of alls	0	4	0
Dec. 27	" 2 pr. of Stuff	0	17	6
1773				
Feb. 2	" 1 " " shoes	1	10	0
	" 1 doz. of Buttons and Nife		6	4
Mar. 6	" 1 silk handkerchief	1	12	0
Apr. 1	" 2 3/4 yds. velvet for jacket	6	17	6
19	" Stuffs for 1 pr. of shoes	0	16	0
	" 24 weeks bord @ 25/	30	0	0
	" cash paid	2	14	0
	Balanced by payment in cash			
		£77	0	0

XII

Newspaper Advertisements for the Shoe Industry

The shoe Industry of New England can be followed in newspapers from the middle of the eighteenth century.

"Gammon Stevens at his shop at North End" offered, among other things, "women's and children's English shoes and galoshes." — *Boston Gazette*, September 3, 1754.

Blanchard Cobb announced that he had imported "women's English clogs and shoes." The following December, on the 8th, Edmund Green, in his shop on Union Street, added to the above list, "children's 1st and 2nd shoes, shoe buckles." — *Boston Gazette*, June 2, 1755.

By 1768, the *Boston Gazette* showed one or more advertisements of supplies of women's English shoes in one paper, as, for example, that of February 15, 1768. Public vendues had become a regular feature of Boston's industrial and mercantile life. Fred Wm. Geyer in an issue of the *Boston Chronicle* for May 23, 1768, offered his imported goods ranging from cinnamon to men's butt soles, French Indigo, and best made Lynn shoes at wholesale or retail, but for cash only. He offered the Lynn shoes by the hundred pair, dozen, or single pair, just as he offered the best G. B. wool-cards by the hogshead, dozen or single pair. By 1769, an advertisement appeared in the *Boston Chronicle* of November 27, of a stock of lately imported men's English shoes and pumps. This is one of the rare mentions of men's shoes in Boston papers.

At Portsmouth, New Hampshire, the *Gazette* for 1757, in its issue of February 23rd, printed an advertisement for Robert Traill, who had English shoes for women, and on May 1, 3, and 20, the same gazette gave the notice that Davenport and Wentworth were advertising their stock of silk clogs, toed clogs, and English shoes.

The supply of imported shoes in Portsmouth was evidently not constant or regular, for, in the issues of the *Portsmouth Gazette* between the above advertisements, while all other kinds of dress and household furnishings are mentioned, no shoes either of domestic or foreign make appear. The domestic shoes, however, were being made by custom makers and perhaps by domestic workers even then up in Portsmouth, for the same Robert Traill, in the issue of the *New Hampshire Gazette* for November 15, 1757, gave the notice that he had received in the Snow Perks from Bristol, England, to be sold at wholesale at his stores on the Long Wharf and his house, all kinds of goods, including shoe-binding galloons, shoemakers tools, knives, hammers, pegging awls and tacks. "I will engage to sell any of the above articles as cheap, for sterling money or Bills, as the like are sold for, anywhere within 250 miles of this Place, and give three, six, and nine Months Credit."

Four years later, in the issues of February 20 and 27, Joseph Barrel advertised at wholesale or retail, women's shoes of all sorts and colours, and three months later he advertised men's shoes. This mention of wholesale and men's shoes is unique at this date. Evidently the custom made or sale shoe has sufficed for the demand and desires of the Portsmouth men until now. The shoemakers of that locality had not been selling at wholesale, nor the stores buying up their product on a large scale for a wholesale trade. Up to this time, all extra or sale shoes had gone into the hands of the general storekeepers of Portsmouth.

Public Vendues of Shoes

Then there came, in 1764, the auction stage, similar to that developed in Boston. On August 3, 1764, and again on August 17, a public vendue at the auction rooms was advertised in the *Portsmouth Gazette* to sell among other articles boots and shoes. "N.B. Goods are received in and sold upon the best terms; secrecy and dispatch observed."

The *Weekly Oracle* of New London, Connecticut, announced on December 31, 1798:

A great variety of ladies and children's shoes, consisting of Black, Red, Green and Straw-coloured Morocco; Fancy and stuff Shoes; Men's shoes . . . for sale by John Lewis.

and added

All those who are indebted to said Lewis by Book [1] or note are requested to call and pay the same by the 15th of January next or they may expect their accounts will be lodged in an attorney's hands for collection.

The expression of "honest as a cobbler" did not always prove true in those days, for the *Massachusetts Spy*, of March 7, 1798, printed a notice of $40 reward:

[1] This refers to the custom of having an account book, where each customer had a page devoted year after year to his account opposite a page which showed the store keeper's indebtedness to him. The accounts were settled at long intervals, and each man signed his name to the pages so that the account book became a receipt book. (See frontispiece for such a case.)

Stolen from the subscriber on the night of the 27th a large black horse ... it is supposed he was taken away by a young lad formerly from Upton by the name of Kelley; he is about 20 years of age, five feet, 9 inches high ... has brown hair, and is a shoe-maker or Cobbler by trade.
Wardsborough, Vermont. JOSEPH WILDER.

Another shoemaker-thief had a pleasing avocation. He was described in the *Salem Gazette* of February 24, 1795:

Stop Thief
On the night of the 25th of January the store of the subscriber was broken open and the following articles taken: — 1 piece of Cassimere, needles and twist. ... There are the strongest marks of suspicion against a person who calls himself Caleb Oaks. He road a dark bay horse which carries his head low, and trots. He is a Shoemaker by profession and *plays on the violin* which he keeps with him. JEREMIAH MAYHEU.

Even forty years earlier there was a dishonest cordwainer about, according to the *Boston Gazette* of August 11, 1755, which printed the following advertisement sent from Charlestown, July 3, 1755:

Charles Raymond, vagrant and thief examined by Thomas Jenness, Justice of Peace, found to be 60 years old and came in 1753 from Bermuda to Connecticut with wife and two children ... and has since been strolling about the country. He is a Cordwainer by trade and has with him 3 pieces of cotton and linen which he confesses he has stolen at Sudbury from a person unknown to him. He was born in Somershire, England, in 1695, and came to Philadelphia in 1710 or 1711.

XIII

NEWSPAPER ADVERTISEMENTS — TANNERIES

The *Boston Gazette* of January 28, 1771 gave the following notice:

To be sold on reasonable Terms, together or in part A Tan-yard, Currying Shop, Bark-house etc. with a dwelling house, barn, and sufficiency of good Land to keep 3 cows and a horse; and in a very convenient place for a Tanner's Business being pleasantly situated in Woburn,[1] at corner of the Roads leading to Andover and Billerica.

Haverhill, May 3, 1799. A Tan-yard, very convenient for a Tanner about half a mile Eastward from Bridge. Whoever has a desire to hire said yard, may have it on reasonable terms by applying to Amos Chase.

[1] Our modern Woburn, Mass., is ten miles from Boston. Even at that early date, the tanning business had begun around the general vicinity of Lawrence and Peabody.

APPENDICES

The *Hampshire Gazette*[1] of April 9, 1788, advertised to be sold

an excellent lot of land lying in Ashfield 2-3/4 miles South of the meeting-house, containing 50 acres on country road leading from Northampton through Ashfield.

Francis Manton who advertised it added, "There is good convenience on the above mentioned lot for a clothier or tanner, both of which are very much wanted." Just a year later, in the same paper,[2] appeared the following:

For Sale, cheap for stock, half a mile east of the Meeting house in Worthington a convenient dwelling house and Barn with a Malt house and Shoemaker's Shop. The buildings well finished with four acres of Land and a Stream sufficient for a Clothier's business or a Tanner.   DAVID WOODS.

One wonders what stock this David Woods wanted. Was it leather to make up into shoes for export trade? Was he about to become an entrepreneur in the shoe industry, or did he intend to go into grazing, or did he want to buy stock [3] in some Western land boom? Probably the citizens of that town knowing David Woods's private affairs sufficiently well, needed no more explanation of this ambiguous mention of payment in stock.

John Chandler when he advertised in the *Worcester Gazette* of May 15, 1794 a Tan-yard farm in Petersham, spoke of it as "well known and well stocked. The farm is too well known to need any encomiums."

The same paper in the issue of November 5, 1794, advertised for sale

A Tan-yard with 13 vats, a Bark House and Mill, a Beam House and a Currier's and a Shoe-maker's Shop with half an acre of Land, situated in Orange. The payment may be made very easy. For further particulars inquire of Sam Farwell of Westminister or Benjamin Wood of Orange.

Hezekiah Hutchins informed his customers in the *Hampshire Gazette* of November 26, 1788, that he had on hand a number of pair of shoes which he would exchange for Wheat, Rye, or Indian corn. He had also a "few Sides of upper and sole leather" which he wanted to exchange for green hides.

May 13, 1795 saw the *Worcester Gazette* offering for sale property which furnished an opportunity for a combination of labor on a farm, in a tan-yard, and a shoeshop.

[1] Published in Springfield, Mass.   [2] *Hampshire Gazette*, for April 15, 1787.
[3] The *Salem Mercury*, of August 14, 1787, and November 27, had been calling attention to the Adventures in the Ohio Company.

A Tan-yard situated on a road leading from Shrewsbury meeting-house to Boylston — also about 20 Acres of valuable land adjoining said Tan-ard with a small dwelling House, Barn, and other necessary buildings which will be sold without the Tan-yard if agreeable to the purchaser, and wi accommodate a Cooper or Shoe-maker. SETH PRA:.

September 30, 1795, another tan-yard was advertised in the *Worcester Gazette*, though the property was in Jaffrey, New Hampshire:

A Tan-yard — one Acre land, small dwelling House and Barn, good ell, a large new Tan House, Beam House, and Currying Shop, in good reair, well situated; being about 100 rds. east of the meeting house in Jaffre on the road from Keene through New Ipswich to Boston and is as good a snd for the tanning business as there is in any country town at that distace from the market as there is no other tan-yard in town that does any gat business; there has been about 200 hides and about 4 to 500 skins tanne in said yard in season for several years. JOHN CUTTE.

One advertiser in the *Worcester Gazette* of September 30, 1795, :r-tainly understood the good points of his property; also the need of markets, and the influence of rehearsing past successes. Anoter would-be seller, Ariel Lumbard, advertising in the *Massachusetts Sy* for March 7, 1798, makes his property so attractive that one woncrs why he had to try to sell it.

A New well situated and convenient Tan Yard about 40 rds west of ie Meeting House in Granby, in the County of Hampshire and within 5 mileof the Canal, with 2 Acres of land, a House, Bark House, and a Shop conveniit for a Currier, Shoemaker, or Saddler, and 10 Tan Vats; together wit a large supply of Hides and Bark, if applied for soon.

XIV

EXEMPTION FOR SHOEMAKERS DURING THE REVOLUTION

STATE OF NEW YORK,
In Senate, March 2d, 1770.

The Senate being informed that the hides which the Convention f this State sometime ago put into the hands of Messrs Matthew Cantie and John Anthony at Marbletown to be tanned and droped (?) b them for the use of this State or some considerable part of them a: prepared for working up into shoes.

Resolved, if the honorable House of Assembly concur herein, Tht Colonel Peter T. Custonius the Commissary appointed to procu Cloathing for the Troops raised under the Direction of this State, tak

he said Quantity of Leather into his Care and cause the same to be
made up into Shoes with all the possible Dispatch to be delivered by
him or his Order into the Cloathing Stores of this State: And that Mr.
Curtenius be & hereby is authorized to give Exemption from Militia
Duty to such shoemakers their Journeymen & Apprentices as he shall
employ in making the said Shoes; to avail them respectively no longer
than during the time they shall severally be in the said employ.
Ordered that Mr. Roosevelt carry a copy of the aforegoing Resolution
.p the Honble —

New York in the Revolution as Colony and State, p. 132. Records discovered & arranged by James A. Roberts, 1897. Pub. at Albany, N. Y.,
1897.

XV

Excerpts from the Wendell Papers [1]

Portsmouth Sept. 8, 1802.

Mr. John Welsh jun^r

Dear Sir

My Son who keeps that shoe store which belonged to our late Friend Mr. Abner Newhall brought me y^r letter of 19th July by which I find that . . . called on him for a Payment of 76 lbs. Ballance — Our poor Friend died the 12th of August of a fit at the store after a short illness and by his Death his Friends are deprived of an amiable man who had he have lived w^d have proved an excellent member of society. His Father came too late to see him alive. As I have a knowledge of his affairs and that by a prudent management he will pay all his Debts and leave 2000 Dollars, I have persuaded his Father to take Administration and for his encouragement I have consented that my Son Jacob shall assist him under the direction of Mr. John Badger Tinman and a neighbor to the deceased. My Quitting Business long ago made me decline taking any charge of his affairs but as an Adviser.

The Pleasure I had of a short acquaintance with you and the Possibility of some future Connections with my Son Jacob who tho young has obtained the character of a very promising and surprising genius. He has the care and division of the whole Stock in trade of the deceased and the Sales go on as rapid as ever. I expect old Mr. Newhall

---

[1] These are in the possession of John Wendell's grandson, Prof. Barrett Wendell of Harvard University.

will be here soon and I shall persuade him to pay the creditors off as fast as is prudent. The law allows him a year to look into his affairs but as I well remember that Abner gave me all his letters from you and that your Boot Legs came to him on very reasonable Terms, I shall influence the old Gentleman to pay you at least 50 Dollars; and as I find my Son proves so very attractive to his Business I have thought proper to encourage him and for that purpose, if you wish to continue the shipping of y^r Manufactory and will ship to me 200 Dollar worth such as you sent to Mr. Newhall, my Son shall sell them and if you want Sole leather shipped you or anything else or if you wish to procure a quantity of Barley for your Market or Bills of Exchange I will introduce him to y^r Business and as he is under age I will be answerable for his Honesty and Faithfulness. Mr. Badger will write you about your debt but he can do nothing without old Mr. Newhall's consent and I will put in a word to him.   I am with Esteem

Your Friend and (etc)

JOHN WENDELL,

claim *vs.* ESTATE of A. NEWHALL $76.16

Coppy of Letter to John Welsh Jr.
Sept, 8, 1802.

Evidently this John Wendell had had a general store (previous to this correspondence of 1802–03) where he sold shoes, among other merchandise, for this page appears:

JOHN WENDELL ESQ. 1798 ACCT. WITH HIS SON GEORGE W. WENDELL.

1797	Debtor		£	s.	d.
Aug. 9	To 6 pair shoes @ 5/pair		1	10	
Mar. 7	" 16 pr. men's shoes @ 5 / pair		4		
	" 1 " new boots		1	4	
	Credits his son by 1 pr. of boots returned		1	4	
Mar. 24	" by 3 pair men's shoes 5 /			15	5
Aug. 20	" " shoes for boys			8	3

ABNER NEWHALL'S ACCT SETTLED JUNE 23, 1800

1800	JOHN WENDEL SQUIRE TO ABNER NEWHALL DR.	D.	C.
6M 2th	To cash	20	0
9th	To cash	40	0
14th	To cash	60	0
18th	To cash	20	0
11th	To 1 pr. of morroco Shoes	1	25
"	" " " " men's "	1	71
23rd	To cash	44	74

"D. is sign he used for dollars
C. " " " " " cents

The next letter shows how Mr. Wendell got into the shoe business:

PORTSMOUTH, Dec. 23, 1802.

Mr. JOHN WELSH Junr.

Dear Sir:

Since my last I am without your favor. I wrote you in fav etc. . . . I was Bondsman for our late friend Abner for 1000 dollars to Hammon & Co of Boston and finding that if the utmost attention is not paid to the sales of his Stock left on hand his Estate would be greatly insufficient to pay his debts and as I was so largely involved for him, I went to Lynn to see the old Gentleman his father but could not get him to secure me. I thereupon insisted that he should do something or deliver the Property into my hand which he has agreed to so I was his Bondman for his Administration, and in Consequence the property has been delivered to me and my son Jacob Wendell has now the whole charge of the Store under me and Mr. Badger having full Business of his own wishes not the Boot Legs sent upon his Acct unless shipped already in which Case he will be responsible and as he is a stranger to you I will see you secured under those circumstances you will please to ship 200 pr. of the best of your boot legs addressed to me on my acct. and risk to Boston to Mr. Bradley to reship me. I please myself with opening an extensive profitable business with you in the flour and leather business as I have sons well calculated to ingross that employment whatever terms you ship your articles on shall be complied with immediately. Boot legs if sent soon will be good articles for sale. Suppose I was to ship 200 dollars of Shoes will they be good remittance? If my life is preserved I will make it Worthy your attention to Ship largely in future. My daughter Dolly is Very genteely settled with the Gentleman to whom she was engaged, Captain Randall of a fine East India Copper Bottom Ship and is going out on another Voyage to India. After which I doubt not he will retire from Trade. Mrs. Wendell and I are well and all join me in our Respectfull Compliments to yourself and Pardner in Trade or your Pardner in matrimony if that has taken Place. I am with Great Sincerity your devoted and assured Friend. JOHN WENDELL.

LYNN the 4mo 6— 1803.

Respecked Friend

I Received thy letter haveing Dat the 3mo 22 which informed me of your welfare and these few lines is to informe thee

that we enjoye our helths. Thee wrote in thy letter that thee wish for
the Creditors to Come and take shoes. I Dont hold it worth while to
say anything to them about for those that have not took anything
always have declined and I Expect they remain the same. Thee rote
nothing about my Shoes. I wish for thee to make the best of them an
turn them in to Cash as Quick as thee Can for I am in scant of Cash
very much my Do is now $113.34. I have not had my part of money
that thee will see by Compareing the a Compts together. My love to
thee and thy family. Lydia's love to thee in particular our little gerl
goes nicely.

Ly—— Newhall's Letter from Lynn.

PORTSMOUTH, March 24th, 1803.

Messrs. JENKINS and WELSH.

Gentlemen

Your fav of 7th Jan 7 I received and observed the Contents. I wrote you a line to Capt. Ayers bound to S. Carolina and Wilmington with two trunks of shoes which they were to sell or return if they sold they were to send the money to you for my acct. I wrote you M. Badger had received the Boot Legs and Jacob has them selling at his store and I expect will make some remittances as I suppose you gave him some directions what to do I have been looking out for the Shoes, I wrote for but is no damage because Badgers are not all sold — I have told my sons customers that I expect some extra legs from Baltimore, the Season of selling is not so good as the late Season and there happens to be a large quantity of them sent here from Boston and Salem.

By Capt. Edmund Fernald I have sent two Trunks of shoes for Sale and if he sells I shall instruct him to remit you the money or ship the two Trunks to your address at Baltimore from Norfolk. They are charged at the first cost in Lynn without additional cost or Freight if you realize the first cost it is as much as I expect. They are of the best quality of Stuff Shoes. I have paid off a number of the Creditors of our friend Mr. Newhall in shoes at their first cost rather than wait and if the estate proves insufficient they are to allow back the deficiency and if you have a mind you may take your Ballance the same way. Mr. Welsh may be assured that I shall do my utmost to save him whole. The insolvency (if any) will arise from a number of Bad Debts and losses on advantures the Way that I am connected with the estate is by being my friend's Bondsman for $1000 and I am under a necessity of keeping it from eventually being so — the demands are nearly $1000 to my very great surprise . . . . etc.

Your devoted Friend
JOHN WENDELL.

ACCOUNT OF STOCK IN JACOB[1] STORE TAKEN APRIL 1, 1803

234 pair Black Morocco @ 6/	$234.50
201 do Col^d ditto 7/	234.50[2]
22 do Lamb Skin kid 6/	22.00
90 do American ditto 7/6	112.50
187 do Stiff slippers 4/	124.67
236 do of ditto 4/	157.33
90 do of Misses ditto 4/0	60.00
76 do of Misses kid 4/6	57.00
56 do English Mor° and kid 8/6	79.33
44 do English kid 10/6	77.00
32 do Misses Mor° 2/6	13.33
7 do calf skin slippers 4/6	5.25
13 do do do ditto with heels 5/3	11.37
23 do do do ditto with heels 4/	15.33
18 pair Misses calf skin 3/9	11.25
16 do Mens Mor° slippers 7/6	20.00
11 do Boys calf skin shoes 5/	9.17
8 do Boys calf skin Pumps 5/3	7.00
8 do Boys course Shoes 4/	5.33
17 do Mens calf skin 8/	22.66
14 do Mens fine ditto 9/6	22.17
65 do children Mor° and calf skin 2/	21.66
13½ skins Mor° Red and Green 10/	22.50
7 sheep skins 3/	3.50

(Some items omitted)

The whole stock totalled $1528.25 worth of shoes and leather.

Wendell as Middleman Retailer

Invoice of 10 Boxes of Shoes and Boots left in the Hands of Jacob Wendell of Portsmouth . . . for Sales on Commission on Actt. and Risque of Mr. Wm. Rose, merchant at Lynn in Mass. via.

---

[1] Jacob Wendell, son of John Wendell, had taken over Abner Newhall's store, and was running it by and with the advice of his father. This record gives a good idea of the volume of shoe business which had been the outcome of the bondsman-entrepreneur-shoe-merchant business affairs of John Wendell.

[2] The shilling was reckoned here as 16 2/3 cents. 201 pr. at 7 (7 sh) = $234.50.

Box 1 contains	140 pr.	Black Morocco Slip. Nos.	1 @ 5/6	$128.33
2 "	130 pair	" " " "	@ 5/6	119.16
4 "	125 "	" " " "	@ 5/6	114.58
5 "	65 "	Mens fine shoes	@ 7/	75.84
5 "	12 "	Black Heels M° Slip.	@ 6/6	13.00
6 "	31 "	ditto	@ /6/	33.58

13 boxes in all for 1077 pairs in all totaling $1099.45

The above shoes I have left with Jacob Wendell to be sold and accounted for agreeable to his receipt given me this day, July 1, 1804.

Signed   W<small>M</small>. R<small>OSE</small>.

Letters from domestic workers as far away as Haverhill have been saved.

H<small>AVERHILL</small> Dec. 5, 1804.

Mr. W<small>ENDALL</small>

    Sir   I receivd your Ltter and I will *make the boots* so you can have them the first next week and will send them by Jacob I did not see Mr. Done but I left the letter and I shall send you some shoes and Boots as soon as Possible — this from your friend and humble servant

S<small>AMUEL</small> E<small>MERY</small>.

H<small>AVERHILL</small> Dec. 15, 1804.

Mr. W<small>ENDALL</small> Sir —

            I have sent you 20 Mens calf skin shoes which you are to sell for me and it is agreed that you shall have for your trouble all that you can get above 7/6 for and either to return the . . .

from your friend
S<small>AMUEL</small> E<small>MERY</small>.

   20 pair calfskin shoes @ 7/6 $25.00

P.S. I have sent a pair slippers in the Bundle and I wish you to deliver them to Mr. Nalty Pitman and if you want any more, send by Jacob and I will send them.   S. E<small>MERY</small>.

I shall send you some Boots next week and I shall be over in 6 weeks.

APPENDICES

	John Wendell Esqr to Joseph Moulton Jr.		Dr.
June 12th	To 2 New Shoes and to toeing 2	6/3	$1.04
Sept. 4	To 2 New Shoes and setting 2	5/6	.92
Aug. 16	To settling 2 shoes	1/6	.25
1805			
June 7	To 3 shoes and setting one	7/9	1.27
Sept. 18	To 3 New Shoes 2 old 1 toeing 2	12/3	2.04
Nov. 1	To 1 Shoe and Setting 4	5/	.84
1806			
Feb. 28	To 1 new steel toed shoe	2/9	.46
May 20	To 3 New Shoes and Setting 4	9/	1.50
1808			
Jany 8	To corking 4 shoes and setting etc	6/	1.00

Portsmouth, 22 Jan. 1805   Rec'd Payment in full
by Jacob Wendell   Joseph Moulton, Jr.

Interesting evidence of business methods can be gleaned from these Wendell papers.

Portsmouth Dec. 17, 1803.

Account Sales of 85 Pair Boot Legs Sold for Acct. of Mr. John Badger at half profits the same being part of an invoice received from Mr. John Welch of Baltimore on said Badger's account.

By account rendered of said 85 Pairs of boot legs amounting to	$102.18
The above account 1803 to Mr. John Badger for the first cost of the Dec. Boot Legs being 85 pair on average @ 90 cts.	76.50
To Jacob Wendell for his half profit	12.84
To John Badger for his half profit	12.84

Mr. John Badger in acct wth Jacob Wendell

To 2 pair Boot legs Mr. Smith and (?)	2.58	By the first cost of 85 pr. boot legs as above	76.50
To Robert Mindun 5 pair in acct. Sales	6.33	By your half proffets	12.84
To cash paid John Wendell Esq. and 2 pair boot legs to ditto	78.44 2.50		$89.34
Dol.	$89.85		

Retail Trade — Typical Bills

Mr. Samuel Sprague
        To Jacob Wendell   Dr.
1802
Dec. 15   To 1 pair Children Shoes   2/3 $ .38
                         Portsmouth 1804.
                           Rec'd Payment.

Messrs. N. S. and W. Pierce
        To Jacob Wendell   Dr.
1802
Jan. 29   To 1 pair men's Shoes   9/   1.50
Feb. 12   To 1 pair M⁰ Slip. del^d per order   7/6   1.25
Apr. 2   To 1 pair Mens Calf Skin Shoes del^d (Kennard) 9/   1.50
                     Portsmouth Jan. 28, 1804

Jacob Wendell
        Bt of James Goodrich
10 pair Mens Calf skins shoes   @ 7/ $11.67
15 pr. boys shoes Calf skin   @ 4/   11.25
12 pair ditto thick Shoes   @ 3/9   7.50
 2 pair Mens Calf Skin Common   @ 6/   2.00
        Rec'd Payment by
                Jacob Wendell
                      James Goodrich.

Invoice of Sundry Shoes Rec^d of Saml Emery to sell for him and if not sold to be returned on demand.
                   Portsmouth, Dec. 17, 1804.
        20 pair Men's shoes @ 7/6   25.00

Another set of pages in a small account book are headed *Retail*: —. They give an account of sales of shoes sold by Jacob Wendell for James Goodrich and Moses Parker consisting of men's calf shoes and boys' thick calf, etc. These were generally sold one pair at a time for prices varying according to the shoes from 92¢ to $1.75 and generally for cash.

XVI

Excerpts from the Batcheller Papers

To Oliver Ward [1] belongs the honor of starting the manufacture of sale shoes in North Brookfield; and his was the earliest establishment of the kind west of Worcester. . . . He learned the tanner's trade of Charles Brown of Grafton; came from Grafton to North Brookfield a little before 1810 and for a short time carried on the tanning business in Spunky Hollow. He started a shoe manufactory here in 1810, depending mainly on the Southern market for sales of his goods.

[1] Temple: History of North Brookfield, pp. 269–272.

Tyler Batcheller, who had learned the trade in Grafton, worked as journeyman for Mr. Ward for eight years, living in his family until 1819. Mr. Ezra Batcheller, a younger brother, also learned the trade of shoemaking at Mr. Ward's, living in the latter's family for six years. In 1819, Tyler Batcheller [1] commenced business on his own account at the Wetherbee house, where he lived with his family, the back part of the house serving as his manufactory. At first his entire business consisted only in what shoes he could make with his own hands; soon, however, he took into his service one or two apprentices and his brother Ezra.

The private expense account of Tyler Batcheller before he became an entrepreneur and began to manufacture on his own capital in 1819, is a wonderfully vivid illustration of the saying that "Capital arises slowly and solely from savings devoted to the production of wealth." For the first three [2] years of his service as apprentice with Mr. Oliver Ward, Tyler Batcheller was still in his minority and his stipulated wages went to his father. Over and above this, during those seven years he earned and saved five hundred dollars, the interest of which was his self-restricted annual allowance for clothing for several years until he went into business on his own account in 1810.

. The first shoes they made were chiefly of a low priced quality especially adapted to the Southern trade. . . . In 1821, he purchased the ' Skerry House ' and farm in the centre of what is now the main village of the town of Brookfield.[3] Here he continued his business in an outbuilding on the premises for three years. In 1824, having taken into his service several additional employes, he built a small two-story shop which is now a part of his immense structure, known far and wide as the 'Big Shop,' into which, on January 1, 1825, he removed his business and took his brother Ezra as partner, under the firm name of T. & E. Batcheller. From this time forward to the end of Tyler's life, the two brothers were associated as partners. Tyler attended to the purchase of stock and to all other business abroad, while Ezra was the efficient and popular superintendent,[4] almost always at home and at his

[1] Temple: History of North Brookfield, p. 270.
[2] Ibid., p. 273.
[3] This is where the large new Batcheller factories, famous in the 60's and 70's, stood.
[4] This account of the Batcheller Brothers, written for Temple, the author of the History of North Brookfield, by Charles Adams who was bookkeeper for the Batchellers for over a score of years, has a personal atmosphere that seems worth preserving.

post, giving direction to all matters pertaining to the manufactory. ...
"The Deacon (meaning Tyler) and Ezra" was the general name for the
brother partners.

XVII

PAGES FROM THE BATCHELLER ACCOUNTS [1]

An account book, called Ledger A, contains all the business transactions of the Batcheller firm from January 1, 1830 to January 6, 1834. Sample pages are given here to show the volume, prices and organization of the business.

Suggestions of the leather and other supplies, bought 1830–31.

DR. 1830				P. R. SOUTHWICK 1830				CR.
Sept.	4	To Sundries	277	$1112.44	Aug. 14	By leather	230	$651.64
					" 25	" "	243	460.80
				1112.44				1112.44
Oct.	4	To Draft	301	258.86	Oct. 4	By leather	301	258.86
"	20	" Note	327	117.75		" "		117.75
"	28	" Drafts	338	796.68		" "		796.68
				1173.29				1173.29
Nov.	29	To cash	29	527.74	Nov. 10	" "	7	259.67
					" 29	" "	29	268.07
				527.74				527.74

Ledger pages show that the T. & E. Batcheller firm bought following goods from various concerns:

Leather from Jones, Wood & Co. — Enfield, Conn.
Leather from Merrick and Dawley — Worcester, Mass.
Shoes from Oliver Ward — N. Brookfield, Mass.
Store [2] supplies (Blankets, cloth, etc.) from Newhall & Co., — West Brookfield, Mass.
Boxes from Elijah Bates — N. Brookfield, Mass.
Lasts from Harrington (Sam). — New Braintree, Mass.
Thread (large quantities), from Nath. Faxon. — Boston, Mass.
Boots & Leather from Thomas Pierce, — Spencer, Mass.
Nails from James Butler, — Boston, Mass.
Twine from Allen Pratt, — Boston, Mass.

[1] The account books of the Batcheller firm covering the period from 1830 to 1895 are in the possession of Mr. Francis Batcheller of North Brookfield.

[2] Evidently the Batchellers ran a general store, just as the Littlefields did, in East Stoughton.

APPENDICES

CUSTOMERS AND THEIR DISTRIBUTION, 1830

(Taken just as the accounts follow in the ledger)

Joseph Whitney & Co.,	Boston, Mass.
Wilson L. Becktell & Co.,	Philadelphia, Pa.
Stoddard Davis,	Charleston, S. C.
Silas D. Edson,	Alexandria, Louisiana
Merriam & Broaddus,	" "
Robinson & Olds,	New York [1]
Aaron D. Harmon,	Alexandria, La.
Asahel Weston,	Baltimore, Md.
Goble & Thomas,	Newark, N. J.
Walter Clarke,	Washington, D. C.

Customers 1830–31.

Alden & Co.	Philadelphia, Pa.
Shipman, Robinson & Co.	Newark, N. J.
Luke Reed & Co.	Augusta, Ga.
Joshua C. Oliver & Co.	Philadelphia, Pa.
John Albree & Co.	Pittsburgh, Pa.
Cheever & Stafford,	Stocking, Yates Co. N. Y.
Mitchell & Bryant,	Boston, Mass.
C. Newhall & Co.	Boston, Mass.
Gage, Stevens & Co.	Cincinnati, Ohio.
Knower & Winslow,	Roxbury, Mass.
Gibbs & Coyle,	Washington, D. C.
N. E. Bank	Boston, Mass.
Nath. Cooper	Charleston, S. C.

New Customers 1831–32.

Shaw, Tiffany & Co.	Baltimore, Md.
J. J. Hammett,	Pittsburgh, Pa.
Bent & Wyman,	Philadelphia, Pa.
Jones & Woodward	Baltimore, Md.
A. P. Childs,	Pittsburgh, Pa.
Ezekel Wood,	Savannah, Ga.
Jacob W. Weaver,	Louisville, Ky.
Otis Spear & Co.	Baltimore, Md.
Griffin Stedman,	Hartford, Conn.
Halsey & Utter,	Newark, N. J.
Abraham Skinner,	S. Brookfield, Mass.

[1] This firm bought on March 6, 1830, $754.80 worth of shoes.

New Customers 1832–33.

Calvin How & Co.	New York City, N. Y.
Amos K. Smith	New Salem, Mass.
C. B. Granniss & Co.	New York City, N. Y.
C. R. & T. J. Comstock,	Hartford, Conn.
Adam Lee & Co.	Rahway, N. J.
Christopher Huntington,	Hartford, Conn.
A. Wood & Company	Savannah, Ga.
Gideon Lee & Co.	New York, N. Y.
Benj. Abbott,	Providence, R. I.
Moses Dickinson,	Detroit, Mich.
Squier & Ross,	Rahway, N. J.
Thomas R. Brooks	Baltimore, Md.
Amasa Walker	Boston, Mass.
C. R. & A. Stone,	Shrewsbury, Mass.
Eveleth & Wood,	Boston, Mass.
George W. Holland,	New York City, N. Y.
John Haseltine & Co.	Philadelphia, Pa.
Woods & Wright,	New York City, N. Y.
Spear & Pattin,	New York City, N. Y.

SAMPLES OF ACCOUNTS.

Batcheller Papers, Ledger A, p. 303, G. W. Holland as customer, 1830.

DR.   CR.

1830   George W. Holland, N. Y.

1830					1830			
Feb.	6	to shoes		96.10	Feb.	6	1830 by leather	158.78
"	16	" "		51.30	"	16	" check T & E. B.	33.92
Mar.	3	" "		48.60	Apr.	5	" draft	207.30
Apr.	5	" "		204.00				
				400.00				400.00

1830					1830			
Apr.	6	To shoes		210.00	June	26	By accpt.	285.00
	24	" "		75.00	July	9	" cash	249.50
June	19	" "		499.	"	19	" Sunds.	1011.50
July	10	" "		762.				
				1546.00				1546.00

1830								
July	20	To shoes		811.50	Aug.	9	By cash	811.50
Aug.	7	" "		144.16	Sep.	21	" "	1389.52
	9	" "		811.50	"	21	" note	552.90
	16	" "		959.00	"	21	" cash	303.36
	23	" "		1547.50	"	21	" "	411.84
						22	" notes	805.40

APPENDICES

Ledger A, p. 306, Spear & Pattin, N.Y.

Dr.		Spear & Pattin, N.Y.					Cr.
1830				1830			
Feb.	6	To shoes	110.75	July	31	By transfer	110.75
Sep.	21	" "	368.60	Oct.	22	" shoes	6.00
"	29	" boots	196.88	Nov.	2	" boots	9.37
Oct.	8	" shoes	191.25	Dec.	28	" drafts	983.01
"	20	" "	50.40				
Nov.	2	" "	191.25				
			1109.13				1109.13
1831				1831			
Mar.	4	To shoes	579.00	May	21	by——	889.00
"	14	" "	160.00				
Apr.	16	" "	150.00				
			889.00				889.00
1831				1831			
May	24	To shoes	225.00	Aug.	8	by note	225.00
Aug.	23	" "	712.	Oct.	29	" "	2290.62
Sep.	9	" "	384.				
·	13	" "	170.			balance	219.50
	21	" "	400.				
Oct.	12	" "	617.				
"	17	" "	220.				
"	29	interest	7.12				
			2735.12				2735.12

Ledger A, p. 309. Stoddard & Davis, Charleston, S.C.

Dr.							Cr.
1830				1830			
Feb.	11	To shoes	324.	May	3	by accpt.	399.39
May	8	" sunds.	75.39				
			399.39				399.39
1830				1830			
Aug.	30	to shoes	600.	Nov.	20	by accpt.	600.00
Oct.	26	" "	399.50	Jan.	8,1831	By "	1197.50
Nov.	29	" "	461.50				
"	30	" "	336.50				
			1797.50				1797.50

Ledger A, p. 309. Stoddard & Davis, Charleston, S. C. — continued

Dr.							Cr.
1831				1831 [1]			
Mar. 21	to shoes		240.00	Aug. 8	by note		243.60
" "	" "		1148.75	Sep. 8	" accpt.		1148.75
" "	" int.		3.60				
			————				————
			1392.35				1392.35

Ledger A, p. 321. Luke Reed & Co. Augusta, Ga.

Dr.						Cr.
1830				1830		
Aug. 6	to shoes	836.30		Aug. 6	by draft	836.30
" 16	" "	700.		Sep. 21	" "	1075.
" 30	" "	375.				
		————				————
		1911.30				1911.30
1830				1830		
Oct. 5	to shoes	755.		Dec. 11	by acct.	1529.50
Nov. 16	" "	774.50				
		————				————
		1529.50				1529.50
1831				1831		
June 28	to shoes	961.32		July 30	by accpt.	961.32
July 27	" "	982.50		Oct. 15	" notes	2286.98
Aug. 29	" "	957.50				
Sept. 7	" "	321.87				
Oct. 15	interest	25.11				
		————				————
		3248.30				3248.30

MEN WORKERS

Nathan Snow
Mathew Edmonds } "making shoes"; paid $307, $237, $241 at a time.
Gideon B. Jenks.
Wm. Levering paid $187 at a time.
John Watson paid $220 at a time.
Leonard Stoddard.
Josiah Hunter.
Wm. Jenks.
Chellus Keep.
Otis Daniels.
Luther Holmes.

[1] In 1833 — August 5 to November 20, the same firm bought $3,787.07 worth of shoes.

WOMEN WORKERS [1] (CLOSING AND BINDING BROGAN UPPERS)

Betsy Perry.
Dolly Hubbard.
Mercy Doane.
Nancy How.
Mrs. Wilder.
Maris Edgerton.
Mary Potter,
Anna Spooner.
Hannah Jenks.
Dolly Ayers.
Mercy Wait.
Rhoda Harwood.
Martha Keith.
Mary Green.
Widow Lucy Lane.
Eliza Hooker.
Persis Howe.
Huldah Knight.
Nancy Blake.
Rachael Ayres.

NOTES PAYABLE,[2] 1830–1831.

March 18, 1830 to January 19, 1831, amounted to $20,075.51
February 4, 1831 to April 27, 1831, amounted to 18,284.05

NOTES RECEIVABLE

January 21, 1830 to February 4, 1831 amounted to $27,784.81
"   "   to April 20   "   " 38,172.28

XVIII

THE BARBER STATISTICS [3]

Date 1837 Town	Pop.	Pr. Shoes	Pr. Boots	Value	Males	Females
FRANKLIN COUNTY						
Erving	292	744	2,050	4,345		
Shelburne	1,018			4,000		
Wendell	847			5,250		
HAMPDEN COUNTY						
Brimfield	1,518	36,000	10,000	58,650	125	50
Palmer	1,810			7,956		
Springfield	9,236			16,000	56	
Wales	738	9,053	6,230	27,743	42	5
Wilbraham	1,802			8,498		

[1] These were either spinsters or widows. Probably many other women, wives and daughters, were doing the same work, but the account appears in the name of the husbands who took out the work and collected the pay.

[2] The amount of these notes, payable and receivable, gives an idea of the volume of the business and of the amount of credit an entrepreneur could expect to receive and might be obliged to give.

[3] Based on facts and figures given in Barber: Historical Collections of Massachusetts, pp. 32–634 (scattered throughout the book).

APPENDICES

THE BARBER STATISTICS (*continued*)

Date 1837 Town	Pop.	Pr. Shoes	Pr. Boots	Value	Males	Females
HAMPSHIRE COUNTY						
Enfield	1,058			11,729		
Ware		61,623	867	53,164		
MIDDLESEX COUNTY						
Bedford	858	90,000		50,000	60	80
Billerica	1,498	19,336	512	11,093		
Burlington	522	5,800		4,900	12	9
Cambridge	7,631			28,768	73	
Dracut	1,898	13,600	700	12,000		
Framingham	2,881	34,955	1,524	34,293		
Holliston	1,775	244,578	20,803	241,626	312	149
Hopkinton	2,166	15,600	72,300	152,300	234	24
Lexington	1,622			12,278		
Lowell	18,010	12,350	3,450	27,240	51	19
Malden	2,303	155,800	250	118,410	24	110
Marlborough	2,089	103,000		41,200		
Natick	1,221	250,650		213,052	263	189
Pepperell	1,586	30,000	100	25,000	30	15
Reading	2,144	290,511	707	184,583	338	494
Sherburne	1,037	48,000	40	40,000	60	30
S. Reading	1,488	175,000 ladies		142,000	260	186
Stoneham	932	380,100		184,717	297	180
Stow	1,734	61,044	537	18,905	32	30
Waltham	2,287			17,787		
Wayland	931	29,666	230	22,419		
W. Cambridge		31,000	500	25,500		
Weston	1,051	17,182	5,606			
Woburn	2,643	279,804	800	221,251	383	320
NORFOLK COUNTY						
Bellingham	1,159	220	14,570	28,077		
Braintree	2,237	71,117	65,604	202,363	357	265
Dedham	3,532	18,722	7,175	32,483		
Medway	2,050	100,650	38,494	149,774	198	98
Needham	1,492	22,673		14,964	26	41
Quincy	3,049	18,603	27,437	111,881	163	58
Randolph..........	3,041	470,620	200,175	944,715	804	671
Stoughton	1,993	53,250	174,800	487,390	495	386
Weymouth	3,387	242,083	70,155	427,679	828	519
Wrentham	2,817	150	10,155	18,675		

The Barber Statistics (*continued*)

Date 1837 Town	Pop.	Pr. Shoes	Pr. Boots	Value	Males	Females
Plymouth County						
Abington	3,057	526,208	98,081	746,794	847	470
Bridgewater	2,092	53,800	3,124	57,317	150	56
Duxbury	2,789	42,334	1,000	56,917	61	60
E. Bridgewater	1,926	263,000	15,100	277,800	270	144
Halifax	781	30,600		27,540	40	
Hanover	1,435	12,000		10,500	35	26
Hanson	1,058	48,000		40,000	180	240
Hingham	3,445	5,654	26,064	55,967	71	51
Middleboro	5,005	shoes &	boots not	given		
N. Bridgewater	2,701	22,300	79,000	184,200	750	375
W. Bridgewater	1,145	27,890	2,518	31,200	43	25
Suffolk County						
Boston		24,626	15,047	102,641	304	55
Worcester County						
Athol	1,603	38,333	16,312	58,741	79	37
Bolton	1,685	20,700	100	6,250	27	13
Boylston	821	17,535	1,300	20,000	34	6
Brookfield	2,514	182,400	17,244	190,697	262	215
Charlton	2,469	15,500		13,700	27	78
Dudley	1,415	27,740		22,698	26	18
Fitchburg		no shoe	return			
Grafton	2,910	671,558	18,672	614,141	906	486
Hardwick	1,818	5,000		14,500	20	8
Harvard	1,789	10,000	5,800	20,500		
Hubbardston	1,780	1,100	5,300	14,562		
Mendon	3,657	150	22,225	39,800	61	6
Millbury	2,153	80,500	9,800	93,175	150	63
Milford	1,637		128,000	212,200	305	37
Northborough	1,224	20,800	7,255	31,720	50	25
Northbridge	1,409	53,500	600	50,000	75	20
N. Brookfield	1,509	559,900	24,170	470,316	550	300
Oxford	2,047	33,522	4,165	36,794	66	45
Paxton	619		24,200	48,430	53	9
Princeton	1,267	50,000	10,304	20,000		
Rutland	1,265	5,950	10,304	23,369	37	13
Shrewsbury	1,507	93,101		88,993	140	109
Southborough	1,113	39,312	170	31,560	80	75
Southbridge	1,740	15,475	590	15,712	17	14

THE BARBER STATISTICS (*continued*)

Date 1837 Town	Pop.	Pr. Shoes	Pr. Boots	Value	Males	Females
WORCESTER COUNTY						
Spencer	2,085	2,940	59,091	106,496	162	28
Sturbridge	2,004	12,660	2,220	18,306	35	15
Sutton	2,457	51,968	9,314	55,656	103	99
Templeton	1,690	9,280	8,530	22,327		
Upton	1,451	3,500	117,699	107,796	156	81
Westborough	1,612	120,656	20,092	148,774	360	214
Worcester	7,117	27,075	18,697	59,320	89	33
ESSEX COUNTY						
Andover	3,878			46,500		
Beverly	4,609			60,000		
Boxford	964			52,975		
Bradford	2,275	360,000				
Danvers	4,804	615,000	14,000	435,900	666	411
Haverhill	4,726	1,387,118	12,003	1,005,424	1715	1170
Lynn	9,323	2,543,929	2,220	1,689,793	2631	2554
Lynnfield	674	54,000	100	40,250	93	80
Marblehead	5,549	1,025,824	97	367,780	503	655
Methuen	2,463	211,300		159,225	190	167
Middleton	671	500	300	1,500		
Newburyport	6,741			113,173	206	114
Rowley	1,203	300,250	32,600	315,360		
Salisbury	2,675	65,500		40,800	87	48
Saugus	1,123	190,326		149,847	269	114
Topsfield	1,049	124,396	900	198,676	272	269

Summary of facts from Barber [1] about other industries which competed with the manufacture of boots and shoes.

All *Cape Towns, i.e.,* in Barnstable and Dukes Counties, were devoted to fishing, shipping and salt making.

In *Berkshire County*, where the climate was adapted to grazing and the water power made manufacturing possible, the efforts and capital of nearly every town went into raising Merino and Saxony sheep or into woolen, cotton and paper mills.

In *Bristol County*, where water power and coast line were abundant, the whale fishery, shipbuilding, and shipping was supplemented by the manufacture of iron and tools, cotton and straw hats.

[1] Facts given in Barber: Historical Collections of Massachusetts, pp. 32–634.

APPENDICES

In *Plymouth County*, its eastern neighbor, where there was also a long coast line, the soil was sandy and unproductive. The coast towns even, divided their labor and capital between sea pursuits like fishing and shipping, and boot and shoe manufacture which was the big industry even in 1837 of most of its inland towns.

In *Norfolk County*, on the north of Bristol and Plymouth Counties, there were straw hat manufactories and shovel making, as in Bristol County, and intensive farming in the portion nearer Boston, but the leading industry was like that of its neighbor, Plymouth County, *i.e.*, the manufacture of boots and shoes. In 1837, 5259 persons engaged in it, out of a population of 50,399, *i.e.*, nearly one-tenth. Norfolk County was also a tanning and currying neighborhood just as Middlesex and Essex Counties were. There seems to be the same connection between sheep raising and woolen manufacture in the western counties and between tanneries and shoemaking in the eastern counties. Weymouth,[1] in 1837, beside making 70,155 pairs of boots and 242,083 pairs of shoes (to the value of $427,679), tanned and curried $42,500 of leather.

Directly north in *Middlesex County*, where the "soil varied" and the "manufacture of cotton goods in 1837 was three times the value of cotton manufacture in any other county of Massachusetts," there were twenty-four towns manufacturing either boots or shoes or both. Bedford was a "shoe town" where there were 30 dwellings and of the small population of 858 people, 60 men and 80 women were at work on shoes. Holliston was already known as a shoe town and Hopkinton, its neighbor, a boot town, in 1837. Reading had for its "great staple and settled business" the manufacture of ladies' shoes. Two hundred and fifty out of 400 males in the town were engaged in it. Malden had not only a relatively and absolutely large shoe business for its size, but had also five establishments for currying leather — doing 28,500 sides in 1837. Woburn also had for its chief industries the manufacture of boots and shoes, and tanning.

Waltham and Lowell were exceptions, being then engaged mainly in cotton manufacture. These were balanced by the other relatively large towns of over 2000 population in the country, like Framingham, Marlborough, Malden, Hopkinton, Reading and Woburn, which

[1] Brockton (North Bridgewater) is only a seeming exception to this rule, when it comes to speaking of individual towns. Brockton did not tan much of any leather, but was so near to Weymouth on one side and Boston on the other that the general condition was the same.

manufactured boots and shoes in large quantities, though not to the extent of putting their county of Middlesex in the list of the first four, Essex, Worcester, Plymouth, and Norfolk, in that trade.

In *Essex County* where the soil was poor and commerce, fisheries and shipbuilding were available industries only for the sea coast towns, the tanneries and the shoe business took the lead, and invaded even such seaports as Marblehead, Newburyport and Salisbury. In many of the towns these latter industries were neighbors if not partners. In Danvers,[1] with its population of 4804, the output of shoes in 1837 was 615,000 pairs of boots and 14,000 pairs of shoes, employing 666 men and 414 women, while it had 110 men employed in its 28 tanneries.

Haverhill and Lynn, the two largest shoe centres, had four and six tanneries respectively which were good sized for the times. Even a relatively small town like Rowley, with 1203 population, was running 16 tanneries besides putting out 300,250 pairs of shoes and 32,600 pairs of boots that year. A casual study of the figures of tables quoted above will make the reader realize that Essex County was leading in the output of boots and shoes in Massachusetts in 1837.[2]

In *Franklin County*, where besides sheep raising for wool, and farming, there was the raising of large droves of cattle for market, there were boot and shoemaking for market in only three towns, and no signs of tanning on a scale large enough to be reported. Palm leaf hats and scythes provided work for domestic workers instead of boots and shoes.

Hampden County was like Franklin County in natural resources and in manufactures. Only two towns report manufacture of boots and shoes for export.

In the length and breadth of *Worcester County*, however, where soil was good and farming was the chief industry, there were 30 towns and cities manufacturing boots and shoes. North Brookfield had outstripped any single town in the state.

[1] Lawrence, the largest modern leather manufacturing town, was a part of Danvers then, and so was Peabody.

[2] In 1905, only one city went beyond Lynn in the value of its output of boots and shoes. While Lynn's amounted then to $25,952,571, that of Brockton was $30,073,014.

XIX

PAGES FROM THE HENRY WILSON ACCOUNTS [1]

HENRY WILSON, STOCK BROGANS ON HANDS OF CONSIGNEES — NOT RETURNED.

1846	No.	Size	Selling Price	Total Receipts
Jan. 1	200 prs.	8–14 B.	75c.	$150.00
	210 prs.	9–14 B.	75	150.00
	400 prs.	7–12 B.	70	280.00
	480 prs.	6–12 B.	70	336.00
	650 prs.	6–11 B.	60	390.00
	1100 prs.	1– 5 B.	40	440.00
	650 prs.	9–13 B.	30	195.00
	240 prs.	6–14 B.	65	156.00
	250 prs.	8–14 R.	75	187.50
	100 prs.	9–14 R.	75	75.00
	210 prs.	7–12 R.	65	135.00
	2150 prs.	6–12 R.	67	1440.50
	1100 prs.	6–11 R.	60	690.00
	550 prs.	1– 5 R.	40	220.00
	250 prs.	9–13 R.	30	15.00
	3180 prs.	6–14 R.	65	2000.00
	200 prs.	6–11 R.	65	130.00
	350 prs.	6–11 R.	70	240.00
	100 prs.	8–14 R.	80	80.00
	120 prs.	6–12 R.	70	84.00

1846 STOCK OF BROGANS SOLD NOT SETTLED.

1846	No.	Size	Selling Price	Total Receipts
Jan.	100 prs.			$75.00
	110 prs.			217.50
	300 prs.			25.00
	50 prs.			60.00
	100 prs.			170.00
	200 prs.			192.00
	250 prs.			42.00
	60 prs.			326.25
	450 prs.			348.00
	480 prs.			490.00
	750 prs.			90.00
	150 prs.			418.00

[1] This account book is owned by Mr. Louis Coolidge, the Treasurer of the United Shoe Machinery Company and a direct descendant of Mr. Wilson.

APPENDICES

1846 Stock of Brogans on Hand Unsold.

Jan. 1	150 prs.	9–14 B.	75 c.	$112.50
	150 prs.	9–14 B.	75	112.50
	360 prs.	9–14 B.	70	252.00
	850 prs.	6–11 R.	60	510.00
	1500 prs.	6–11 R.	65	975.00
	60 prs.	6–14 B.	60	36.00
	100 prs.	12–14 B.	90	96.00

Stock in Hands of Workmen.

Jan. 1	900 prs.	6–11 R.	65 c.	$585.00
	50 prs.	9–14 B.	75	37.50
	120 prs.	6–24 B.	60	72.00

Stock on Hand.

Jan.	1000 ft.		11 c.	$110.00
	2000 ft. Kipp L.		10	200.00
	Sole leather,			100.00
	Innersole leather			50.00
	Welt Leather			100.00
	30 doz. long (?)			100.00
	splits			130.00
	11 cases upper L.			100.00
	3 cases upper L.			15.00

1846 Real Estate on Hand.

Jan. 1	House	$1800.00
	Shop	800.00
	House lot	300.00
	House and shop	1300.00
	House lot	900.00
	Our half house	500.00

Personal Property on Hand.

	Furniture	$600.00
	Splitting mill	50.00
	Horse and wagon	86.00
	Small bills	50.00
	Notes on Hand	$3765.76

Notes and Bills to Pay

Jan. 5	Notes Given	$9780.00
	Stock	9780.00
	Notes in market	4300.00
	Bills not settled	3362.00
	Cash borrowed	265.00

Some Individual Accounts

1846 Dr.   W. A. Ransom,   Cr.

Mar. 7	To 250 pr.	6–11	67 c.	$167.50	By note	$315.49
Mar. 13	To 250 pr.	6–11	67 c.	167.50		19.51
				335.00		335.00

1846 Dr.   E. P. Sanford,   Cr.

Feb. 23	To 100 pr.	8–11	87½ c.		Apr. 10 By cash	$442.75
28	To 250 pr.	6–11	65½ c.			
	To 250 pr.	6–11	65½ c.			
	To 50 pr.	8–12	77½ c.			
				442.75		

1845 Dr.   J. F. Collath,   Cr.

Oct. 16	To cash and orders	$245.24	Apr. 11 By Bal. on leather	$72.74
	To cash	5.00	Oct. 16 By work on Let	227.67
	To cash	5.00	1846	
Dec. 3	To cash	25.00	June 5 By cutting	71.10
1846			By Bot. 100 pr. (3) 17½	17.50
Jan. 1	To cash	15.00	By patterns	3.50
	To goods	12.23	By clos. 180 pr 6–14 (3)	5.40
	To 12 N. F.	2.40	By clos. 50 pr. r. (4)	2.00
	To cash	12.00	Apr. 18 By work	175.00
Apr. 18	To goods & Cash	168.74		574.91
		490.61		
Apr. 27	To note	84.30	Apr. 18 By balance	84.30

1846 Dr.			Spears and Vanderhoof,		Cr.	
Feb. 23	To 150 pr.	6–11	67½ c.			
	To 50 pr.	8–13	80		Mar. 23,	
Feb. 28	To 50 pr.	8–13	80		By note	$615.00
	To 200 pr.	6–11	70			
Mar. 3	To 150 pr.	6–11	67½			
	To 50 pr.	8–13	80			
	To 50 pr.	8–13	82½			
12	To 50 pr.	8–13	82½			
	To 100 pr.	6–11	70			
			$615.00			

These brogans were shipped in boxes [1] — no longer in casks or trunks, as in the pioneer days of trade.

1844 Dr.			Gilman Moore,		Cr.	
Oct. 17	To cash	$180.34	Oct. 17	By boxes		$180.34
June 26	To cash	150.00	June 10	" "		156.36
1846			1846	by balance		6.36
Jan. 1	To cash	141.09	Jan. 1	By boxes		134.73
July 1	To cash	92.07	June 11	" "		84.27
			July 1	" "		7.80
						92.07

1846 Dr.			Priest and Co.		Cr.	
June 1	To cash	$20.00	June 1	by boxes		$26.00
		6.00				

Evidently the boxes of brogans were shipped by freight on the new Worcester Railroad.

1845 Dr.			Worcester Railroad Corporation,		Cr.	
May 3	To cash	$13.00	May 3	By balance		$18.89
6	" "	5.89	June 3	By Bill		16.04
June 3	" "	10.00				
	" "	6.04				
Apr. 22	To cash	14.62	Apr. 20	By balance		14.62
May 5	To cash	4.53	May 5	By Bal.		4.53
June 9	To cash	33.01		By carting		33.01

[1] Compare with use of boxes in Lynn.

APPENDICES

Prices paid for work, and the fact that pay was sometimes given in orders on local stores, are shown in pages like the following:

1845	Dr.	Martin Haynes,			Cr.	
June 3	To goods at C & W	$100.00	Oct.	By clos. to date		$124.35
Oct. 16	" " " " " "	50.00	14	" " and work		30.00
	To goods	150.00				154.35
		4.35				

1845	Dr.	Mrs. A. Davis		Cr.	
June 3	To order on Settlement	$6.50	June 3	By work on leather	$6.50

1845	Dr.	Mrs. Gowan		Cr.	
Oct. 23	To orders and cash	$124.00	Oct. 23	By work to date	$124.00

Sometimes workers were paid partly in lasts.[1]

1845	Dr.	L. D. Moody,			Cr.	
Oct. 24	To 12 lasts	$2.16	Oct. 24	By pr. Brogs. $12		$19.10
Dec. 4	To cash	19.20		By work		2.16

1845	Dr.		John O. Wilson			Cr.	
Oct. 24	To 4 lasts	.72	Oct. 24	By Bott. 100 pr. 12 [2]			$12.00
	" 4 R. "	.84	24	" " 240 " 13,			31.20
Apr. 20	To cash	25.00	1846				
30	To cash	20.00	Jan. 5	" " 50 pr. 14			7.00
	To 20 lasts						
	(20)	4.00		" " 100 pr. 10R			10.00
	To 1 barrel						
	flour	6.62		" " 50 " 14 "			7.20
	To cash	5.00		" " 250 " 17 "			42.50
		62.18	May 5	" " 850 " 17 "			144.50
	To amt. for.	192.22					254.40
							62.18
			May 11	By balance			192.22

[1] This buying of lasts by domestic shoemakers, from Henry Wilson as manufacturer, suggests that he was getting them made at a wholesale reduction. The same thing was being done at Howard and French's factory in Randolph.

[2] These figures are prices in cents, representing price paid per pair for bottoming. Evidently these brogans were roughly and quickly pegged, and at the lowest customary price. The price Wilson paid for closing uppers was 3 or 4 cents a pair. Mr. Isaac Felch has told me that the rate for bottoming brogans ran from 18 to 22 cents a pair, and 4 cents a pair for closing the uppers. The machine can do all this work today for 9 to 11 cents a pair.

1845	Dr.		ALONZO NUTE		Cr.	
Oct. 24	To 33 lasts	$6.12	Oct. 24	By Bott. 350 pr. B [1] 14	$49.00
Nov. 20	To cash	100.00		By " 720 pr. R 13	93.60
	To 12 qt. pegs	.48		By " 150 pr. R 12 [2]	18.00
	To cash	6.00		By " 50 pr. R 12	6.00
	To 14 R. lasts goods			By " 300 pr. R 13	39.00
		———		By " 110 pr. B. 14	15.40
		147.91		By " 150 pr. R 17	25.50
Apr. 1	To cash	181.47			
		329.38	Apr 1	By work	329.38

1845	Dr.		H. W. HAMMOND		Cr.
Oct. 24	To lasts	3.42	Oct. 24	By work to date	85.00
Dec. 8	To cash	50.00	1846		
	To 4 lasts (18)	.80	Jan. 5	By 100 pr. B. 8–14 (15)	15.00
1846				By 25 pr. B.6–11 (13)	32.50
Jan. 15	To bill to note	65.00		By 360 pr.B. 6–12 (14)	50.40
	To cash	60.00	May 5	" " " " 6–12 (14)	50.40
	To cash	50.00		" 50 pr. R. 17	8.50
	To order	25.00		By work	83.45
		254.22			325.25
May 5	To balance	73.53		By work	2.50
		327.75			327.75

This year's business alone was interesting enough to make students of industrial history glad that by happy chance the account book chronicling it was saved. It is so like hundreds of other cases in Massachusetts, however, that it would have not warranted the interest taken in Henry Wilson's "ten-foot factory" by thousands of tourists and politicians, if he had not been known later as the "Shoemaker-Vice President."

[1] R. is russet. B. is black.   [2] Twelve is 12 cents a pair.

XX

Excerpts from the Robinson & Co. Papers[1]

Notes and Bills Received in and After Sept. 1848

Date 1848	From	On wh. acct.	Time	When due	Amt.	Remarks
June 3	Down & Ball	Mdse	6 mos.	Dec. 3/6	117.77	Sent for Disc. Oct. 10th
May 29	D. B. Trufant	Leather	4 mos.	Sept. 29	113.48	Sent to Warren Bk. for collection
Aug. 15	Freeman & Childs	Mdse	4 mos.	Dec. 15	162.12	Sent to Warren Bk. for collection
June 21	Graham & Van Voust	Mdse	4 mos.	Oct. 21	131.66	
July 15	W. N. Spinney	Leather	6 mos.	Jan. 13	168.04	

The "Account of Stock" book of 1854 gives the following items and shows how many of the machines so recently invented were owned by the firm:

```
Accounts Rec'ble (amounted to) ................. $18,518.74
Accounts Payable      "       "  .................  11,140.00
Bills Rec'ble ....................................   9,422.06
Bills Payable ....................................   8,861.42
Mdse. as follows:
  Manufacturer's Stock ..........................   1,686.50
  Leather .......................................   4,723.35
  Trimmings .....................................     257.00
  Machine and fixtures, as follows:..............     954.06
    1 sole cutting machine ............  $120.00
    1 pr. scales ......................    20.00
    1 safe ............................   127.81
    1 desk ............................    30.00
    1 splittng machine ................    50.00
    1 stripping machine ...............    55.00
    1 Nichols & Bliss sewing machine ..   125.00
    1 Hunts        "         "   .....     65.00
    1 Howes        "         "   .....     50.00
    1 Punching     "             .....      4.00
    1 stove        "             .....      5.00
      coal on hand ................         25.00
      lasts & patterns ............        250.00
      7 stools ....................          5.25
      basket & duster .............          4.00
      clock & stationery ..........         10.00
      Grindstone ..................          8.00    954.06
```

[1] Preserved in the archives of the Lynn Historical Society.

APPENDICES

XXI

Natick Statistics

The Natick industries in 1854, according to the Massachusetts Census of 1855, were as follows:[1]

	No. of manufactories	No. persons employed	Capital invested	Value of articles made
Shoe Box Manufactories	2	10	$5,500	$19,100
Harness Manufactories	2	3	500	2,100
Cap Manufactories	1	1	25	350
Carriage Manufactories	3	10	1,550	5,900
Pulp for paper "	1	12	16,500	70,425
Cutlery Manufactories	3	3	450	2,000
Baking Manufactories	1	4	1,100	9,110
Shoe filling Manufactories	2	3	50	1,500
Wholesale custom cloth Manufactories	5	11	5,500	30,800
Value of tree nails or ship pins prepared for market				4,136
Value of ship timber sold				1,730
Value of ship plank sold				260
Value of shoes and boots				1,163,808
No. of pairs of shoes made				1,281,295 [2]
No. of boots made				570
No. of males employed				1,070
No. of females employed				497

The volume of the Walcott business alone in Natick was as follows:[3]

Year	Pairs of Brogans	Year	Pairs of Brogans
1835	4,050	1845	40,350
1836	11,000	1846	64,000
1837	8,310	1847	100,010
1838	9,290	1848	84,012
1839	10,350	1849	107,336
1840	8,200	1850	104,222
1841	18,700	1851	112,140
1842	21,830	1852	118,080
1843	25,113	1853	118,140
1844	36,710	1854	97,920

[1] Quoted by Bacon: History of Natick, p. 154.

[2] This is interesting because for the last forty years Natick has been known only as a boot town.

[3] Quoted by Bacon: History of Natick, p. 154.

XXII

Excerpts from the Howard and French Accounts[1]

1845
Jan. 14 Amasa Clark, Dr.

To 12 pr. St. Calf Peg'd boots			$25.50	
2 " calf	" ½ welt boots		3.75	
2 "	kip		3.25	
1 "	calf N. P.		2.63	
1 "	Kip "		2.25	
1 "	Calf sewed		2.75	$40.13

Feb. 8

To 12 pr. Kip pegged boots		22.50	
12 " calf " "		27.80	$50.30

Feb. 14
    To 1 pr. St. Calf Pegged Boots $2.50
Feb. 19
    To 12 pr. St. Calf Pegged " 26.50
Feb. 22

To 12 pr. St. " " "		25.50	
6 " " sewed "		19.50[2]	

Mar. 11

12 pr. Goat 10/6[3]	21.00
12 " light Calf	24.00
12 " St. Calf	33.00

Mar. 17 Amasa Clark, Cr.
    By cash 10 dollars 10.00

Mar. 18

24 pr. St. Calf Pegged Boots-wide	11/9	47.00
12 " " " " " "	12/–	24.00
12 " St. Goat Boots	12/9	25.50
12 " " " "	12/–	24.00

Mar. 28

To 12 pr. St. Calf. Peg'd boots	12/9	25.50
12 " " " " wide		24.00
12 " " sewed		37.50

Apr. 1

12 pr. Goat Peg'd Boots (wide)	24.00
12 " " " Best	25.50
12 " ½ W. wide	21.00

[1] These Howard and French account books are in the possession of the author.
[2] Note difference in price between pegged and sewed boots of same stock, $25.50 against $39.00 a dozen pairs.
[3] 10/6 means sizes 10 to 6.

AMASA CLARK'S SALES LARGER IN FALL OF THAT SAME YEAR, 1845.

Sept. 12	To 12 pr.	Calf. Sewed Boots (B. I. & Co.)					$30.00	
	48 "	St. Calf. Pegged Boots					103.00	
	36 "	"	"	sewed			100.50	
	36 "	"	"	"			88.56	
	36 "	"	"	Grd[1]			76.50	
	12 "	"	"	"	DS. Pegged		27.00	
	12 "	"	"	"	"	W. P.	30.00	
	12 "	"	"	Nap.			27.00	
	12 "	"	"	"	D. O.		27.60	$510.16
Sept. 23	To 12 pr.	Calf Pegged ½ W[2]					25.50	
	" "	" " "					25.50	51.00
Sept. 26	To 12 pr.	Kip[3] Pegged W. P.[4] Boots 12/9					25.50	
	12 "	"	"		"	10/6	21.00	
	8 "	Calf	"		"	15/	20.00	
	7 "	St. "	"		"	12/9	14.88	
	3 "	Boys	"	thick	"	8/	4.00	
	6 "	Chil.	"	"	"	6/6	5.50	90.88

Clark collected payment in form of notes and sometimes cash from the customers; the following entries credit him for it:

1845	Amasa Clark, Cr.			
Sept. 26	By J. Richardson	note Apr. 1, 6 mos.	$269.00	
	D. & Bushwell	" Apr. 15, 6 mos.	192.55	
	D. & H. Dixey,	" Mar. 8, 3 mos.	50.00	
	Cash		168.45	
	Total			680.00
May 22	A. Clark Cr. By Cash		125.00	
Jan. 16, 1846	Amasa Clark, Cr.			
	By T. Foss Note Jan. 10, 6 mos.		141.00	
	J. J. Ashby, Bal. on note		34.50	
	John Perly Note Jan. 15, 6 Mos.		200.50	
	Cash Two Hundred Fifty dollars		250.00	
	Discount on J. Richardson Apr.		4.00	
	" " " Kip boots		1.50	
	" " " calf		.25	
	Premium on H. E. S. & CO. goods		11.00	642.75

[1] Grained.
[2] 1/2 welt.
[3] Kip — old calf.
[4] W. P. means water proof.

APPENDICES

Customers and Sales.

1845

Jan. 13[1]	Wm. Dyke, Dr.		
	To 12 pr. Men's Kip W. P. pegged boots		$22.50
	Amasa Clark, Dr.		
	To 19 pr. boots		40.13
Jan. 14	George V. Edney		
	To 6 pr. St. Calf Pegged boots 9, 10, 14/......		14.00
	6 " " Kip " " " 12/......		12.00
Jan. 15	O. Ames		
	To 8 prs. Mens Thk Sewed Boots		17.33
Jan. 18	Caleb S. Small, Dr.		
	To 12 pr. Boys Thick pegged Boots	$17.00	
	12 " Youths " " "	13.20	30.20
Jan. 22.	Joseph Crocker, Dr.		
	To 12 pr. Calf Top Soled Boots		28.50
Jan. 31	Rufus Wade, Dr.		
	To 12 pr. M. Calf Sewed Boots	36.00	
	12 " " " Pegged " ½ W	27.00	
	12 " St. " " "	27.00	90.00
Feb. 8	Amasa Clark, Dr.		
	To 12 pr. Kip Pegged boots	22.50	
	" 12 " Calf " "	27.80	50.30
	William Dyke, Dr.		
	To 12 pr. Mens Kip Pegged W. P. Boots		25.00
	E. Holden, Dr.		
	To 12 pr. Mens Calf Sewed Boots ....;....		31.50
	D. W. Wisell, Dr.		
	To 12 pr. Mens Calf Peg'd W. P. Boots		28.50[2]
Feb. 12	T. P. Edney, Dr.		
	To 12 pr. St. Calf Peg'd boots		27.00
Feb. 14	Wm. H. Learned, Dr.		
	To 7 pr. Men's kip peg'd W. P. Boots		14.58
Feb. 15	D. O. Donnell, Dr.		
	To 50 pr. Boys Kip Brogans		37.50
Feb. 18	George P. Smith, Dr.		
	To 25 pr. Mens Kip Brogans	$ .90	22.50
Feb. 19	Amasa Clark, Dr.		
	To 12 pr. St. Calf Peg'd Boots		26.50
	Rufus Wade, Dr.		
	To 12 pr. St. Calf Peg'd Boots		27.00

[1] These dates were taken at random.

[2] Notice prices of these boots compared with those made for California trade later.

CUSTOMERS AND SALES (*continued*)

1845
Feb. 22 Amasa Clark, Dr.
  To 12 pr. St. Calf peg'd Boots C. J. A. $25.50
    6 " " " sewed W. P. A. S. D. 19.50

Feb. 26 Sam. Dummond, Dr.
  To 24 pr. Mens Grd. ½ W. Boots $1.65   39.60
  " 24 " " St. Calf " 1.87½   45.00
  " 12 " Boys " " " 1.37½   16.50
  " 1000 yds Shoe strings 7.50

Mar. 3 Rufus Wade, Dr.
  To 12 pr. Mens Calf Peg'd ½ W Boots 27.00

Mar. 4 Allman & Maxwell, Dr.
  To 12 pr. Mens Calf Sewed Boots 36.00
  " 200 yds. strings 1.50
  " 12 pr. Mens Lt. Calf Peg'd................ 19.50
  " 12 pr. Grd. ½ W " 19.50

Mar. 5 Wm. Pearman, Dr.
  To 12 pr. Lit. Calf Peg'd Boots 1.87½   22.50
    12 " " " " 20.04

Mar. 7 Amasa Clark, Dr.
  To 12 pr. Goat Peg'd Boots 24.00
  " 12 pr. Calf " " 24.00
  " 12 pr. Mens Calf Sewed Boots 37.50
  " 12 pr. " Gr. " " 25.50

Mar. 11 Levi Mann, Dr.
  To 12 pr. Lt. Calf Peg'd Boots 22.50
  " 72 " " Shoulder 114.00
  " 12 " " Seal 18.00
  Amasa Clark
  To 12 pr. Grd. Peg'd ½ W Boots 21.00
    12 " Lt. Calf. 24.00   45.00

  D. O. Donnell, Dr.
  To 12 pr. Mens Calf Boots 27.00
  O. Ames, Dr.
  To 25 pr. Mens Thick Peg'd Shoes 25.00
    " " 28.12   53.12

Mar. 14 Amasa Clark, Dr.
  To 12 pr. St. Calf Sewed Boots 33.00
  " 12 " " " " " 33.00

Mar. 14 Levi Mann, Dr.
  To 12 pr. St. Calf Pegged Boots 11/3 .... 22.50
    12 " Grd. ½ W " 19.20

APPENDICES

1845
CUSTOMERS AND SALES (continued)

Date	Customer / Item	Price	Amount
Mar. 18	D. O. Donnell, Dr.		
	To 12 pr. Seal Peg'd ½ W Boots		$21.00
	Levi Mann, Dr.		
	To 36 pr. Lt. Calf Peg'd Boots		57.00
	Amasa Clark		
	To 24 pr. St. Calf Peg'd Boots wide	11/9	47.00
	12 " " " " " "	12/	24.00
	12 " " Goat " " Best	12/9	25.50
	12 " " " " "	12/	24.00
	Wm. H. Learned, Dr.		
	To 12 pr. Goat Peg'd Boots	11/7	22.50
Mar. 21	R. W. Howes, Dr.		
	To 60 pr. grd. ½ W Peg'd Boots	9/9	97.50
Mar. 22	G. P. Smith, Dr.		
	To 12 pr. Boys Calf ½ W Boots	10/	20.00
Mar. 22	D. O'Donnell, Dr.		
	To 24 pr. Calf. Downings	6/	24.00
Mar. 22	G. P. Smith, Dr.		
	To 12 pr. Goat Peg'd Boots	11/6	23.00
Mar. 22	Wm. Dyke, Dr.		
	To 12 pr. Goat Peg'd Boots	11/6	23.00
Mar. 24	O. Ames, Dr.		
	To 12 pr. Thick Peg'd Brogans	6/	12.00
Mar. 25	Samuel Dummond, Dr.		
	To 36 Pr. St. Calf Peg'd Boots		67.50
	" 1000 yds. shoe strings		7.50
	" 12 pr. Boys St. Calf Peg'd	8/3	16.50
	" 24 " Youths " "	6/9	27.00
Mar. 28	Rufus Wade, Dr.		
	To 12 pr. Lt. Seal Peg'd Boots		20.00
	Wm. Dyke, Dr.		
	To 12 pr. St. Seal Peg'd Boots		18.60
Mar. 28	Amasa Clarke, Dr.		
	To 12 pr. Lt. Calf Peg'd Boots		25.50
	To 12 " " " " " wide		24.00
	To 12 " " " sewed "		37.50

STOCK[1] BOUGHT BY HOWARD & FRENCH FROM JAN. TO MARCH, 1845

Jan. 4	R. McConnell & Co., Cr.	
	By Mer.[2] per bill	$55.20
	G. G. Gove & Co., Cr.	
	By Mer. per bill	47.78

[1] Stock was bought by this as well as other firms then, in relatively small amounts, judged by standards of today. These firms sometimes sold in still smaller quantities to shoemaker-customers, e.g., 25 or 39 lbs. of sole leather.

[2] Mer. stands for merchandise.

STOCK BOUGHT BY HOWARD & FRENCH FROM JAN. TO MARCH, 1845 (*contin'd*)

Jan. 4	Hunt & Cutter Co., Cr.		
	By Mer. per bill		$64.00
	Wm. Mitchell & Co., Cr.		
	By Mer. per bill		48.03
	W. & J. Guild, Cr.		
	By Mer. per bill		26.09
Jan. 11	R. McConnell & Co., Cr.		
	By Mer. per bill		39.12
	Nath. Faxon[1] & Co., Cr.		
	By Mer. per bill		69.33
	A. Thompson & Co., Cr.		
	By Mer. per bill		54.11
	J. Nichols, Cr.		
	By 153 3/4 ft. wax leather	$13	19.98
	34¼ " split "	30	10.27
	120 " Kip "	12	14.40
Jan. 18	J. Nichols, Cr.		
	By 1 doz. Calf skins 42½ to 3/9		26.56
	1 roll splits 33 to 30		9.90
	Tillson & Mitchell, Cr.		
	By Mer. per bill		26.26
Jan. 18	S. & J. Guild, Cr.		
	By Mer. per bill		51.51
	R. M. McConnell & Co., Cr.		
	By 4 doz. calf skins 76 to 60		45.60
	Proctor & Kendall, Cr.		
	By Mer. per bill		154.00
Jan. 25	J. Nichols, Cr.		
	By 156¼ ft. Wax leather		20.32
Feb. 15	By 155 " " "		20.15
	" 126 " Kip		13.86
	" 41 " Calf		25.62
	N. Faxon & Co., Cr.		
	By Mer. per Bill		83.32
Feb. 15	G. G. Gove & Co., Cr.		
	By Mer. per bill		67.57
Feb. 17	J. W. Harris & Co., Cr.		
	By 25 pr. Kip Peg. Brogans[2]		20.83
Feb. 22	N. Faxon & Co., Cr.		
	By sole leather per bill		24.96

[1] This firm of Nathaniel Faxon sold thread to many manufacturers, as various ledgers show.

[2] Howard & French bought brogans instead of making them when they had an order to fill.

STOCK BOUGHT BY HOWARD & FRENCH FROM JAN. TO MARCH, 1845 (*contin'd*)

Mar. 1	O. Ames & Co., Cr.		
	By 12 doz. knives as per bill		$10.90
Mar. 10	A. Thompson & Co., Cr.		
	By Mer. per bill		48.34
	N. Faxon & Co., Cr.		
	By Mer. per Bill...........................		42.00
	R. McConnell & Co., Cr.		
	By Mer. per bill		81.00
	Hunt & Cutler Co.		
	By 2 doz. Goat (Skins)	$16	32.00
	2 doz. " "	9	18.00
	Proctor & Kendall, Cr.		
	By Mer. per bill		141.33

PIECE WAGES AND SPECIALTIES OF DOMESTIC WORKERS

Andrew M. Fritts, in 1846, was both fitting and making[1] boots of various grades, at a piece wage:

1846
Jan. 13	12 pr. calf boots fitted & made		$11.50
23	12 " " " " " "		11.50
Mar. 24	12 " " " Lt[2] "		5.25
Apr. 2	12 " " " " " " " extra		4.50
May 4	12 " " " " " " "		4.50
19	12 " " " " " " "		4.50
26	12 " " " " " " "		4.50
28	12 " " " Boys maker		4.25
June 8	12 " " " " "		4.25
9	12 " " Goat		4.25
17	12 " " Calf fitted and made		5.75
24	12 " " goat made (boy)		4.25
26	12 " " Calf		5.75
July 3	12 " " wide Lt. Calf (boy)		5.00
24	12 " " Kip W. P.[3] made		5.50
28	12 " " Lt. fitted & made		5.50
31	12 " " " " " "		6.50
Aug. 11	12 " " " " " "		6.50
11	12 " " " Kip W. P. made		5.50
22	12 " " " Lt. Calf		6.50
27	12 " " " Kip. W. P.		5.50

[1] Probably his wife did the fitting, but it was credited to and settled for with him.
[2] Lt. — Light weight calf.
[3] W. P. — Waterproof.

228 APPENDICES

Daniel Wilde was making the same kind of pegged boots at a uniform price. He was credited on

1847
Apr. 1 By 12 pr. pg. Boot made $6.50
May 10 " 12 "   "   "   " 6.50
     20 " 12 "   "   "   " 6.50
     24 " 12 "   "   "   " 6.50

and so on until September. This does not represent his full working time, probably, but all the time devoted to his work for the Howard & French firm.

The account of William Stetson shows still other kinds of boots being put on the market by Howard & French. He was credited

1845
Oct. 3 By 12 pr. grd.[1] D. S. $6.00
     11 " 12 " Thick 4.80
     20 " 12 "   " 4.80
     25 " 12 "   " 4.80
Nov. 10 " 12 " long Grd 6.00
     22 " 12 " thick 4.50
Dec. 1 " 12 "   " 4.50
         " 1 " Long .... .50[2]

These accounts have not only suggested piece wages for different kinds of work, but the variety the firm was making. Sizes began to be carefully noted in the accounts for 1847.

Adoniram J. Dyer, in 1847, was credited

Jan. 15 By 12 pr. Boots made 5/- $10.00
     30 " 12 pr.   "   " 5/-   10.00
Feb. 13 " 13 "   "   " .... 5/ & 6/ 11.00
Mar. 4 " 12 "   "   " 5/   10.00
      4 " 3 "   "   " 6/5   3.00
     24 " 1 "   "   " 5/   1.00
Apr. 1 " 12 "   "   " 5/-   10.00
     12 " 7 "   "   " 6.00
     30 " 12 "   "   " 5/   10.00

This price for making was so much higher that it shows better work in higher grade boots. Perhaps they were hand-sewed instead of

---

[1] Grd. — Grained.

[2] Fifty cents was the price for "making" long boots, presumably because they were more awkward to handle.

pegged, although the account does not say so. In the case of Benjamin Dyer, whose account follows, this fact is mentioned:

1847
Jan. 22 By 12 pr. Calf Sewed 4/5 $9.00
Feb. 6   " 12 "   "   " 4/6   9.00
Mar. 4   " 12 "   "   " 4/5   9.00
     8   " 12 "   "   " .... "   9.00
     24   " 12 "   "   " .... "   9.00

and so on down to January, when the bill due Benjamin Dyer was $180.00.

WOMEN'S WAGES AS DOMESTIC WORKERS ON BOOTS
MRS. IRA HOWARD

1854 Mrs. Ira Howard, Cr.
July 22   10½ doz. topt $1.26
     4   "   corded .... .80 [1]
Aug 5   28   "   topt 3.36
     11   "   corded 2.20
   12   8   "   topt .... .96 [2]
     8   "   corded 1.60
   19   5   "   " 1.00
     17   "   topt 2.04
   26   13   "   " 1.56
Sept. 2   15   "   " 1.80
     9   "   corded 1.80
   30   70   "   topt 8.40
     25   "   corded 5.00
Oct. 28   31   "   topt 3.72
     28   "   corded 5.60
     12   "   fitted 1.25 $42.35

An opposite account shows that each day Mrs. Howard returned the work, she was paid in cash odd amounts, not what the work came to, however, sometimes more, sometimes less. It was finally balanced on October 28, 1854.

IRREGULAR AMOUNTS OF WORK

Edwin Howard, from January 7 to April 8, 1854, treed 100 pairs of boots for $7.00 every week. After that he sometimes went up to 108 pair or dropped to 60 pair in the middle of the summer. In 1855, he averaged 108 pairs, week in and week out.

[1] Twenty cents a dozen, uniform price for cording.
[2] Twelve cents a dozen, uniform price for topping.

Sylvanus Pratt's record in 1854 had been so irregular that one wonders whether this irregularity in output was due to Pratt's speed or to different number of hours spent or to the varying amounts of work given to him to do.

1854	Oct.	28	216 prs. treed		1854	Dec.	30	108 prs. treed
	Nov.	4	173		1855	Jan.	6	108
		11	223				13	100
		18	181				20	96
		25	99				21	96
	Dec.	2	58			Feb.	3	86
		9	93				10	90
		16	108				17	94
		23	118					

Henry Bangs, the fifth treer in Howard & French's shop, was even more irregular in amount of work done a week, varying from 48 to 112 pairs a week.

XXIII

Excerpts from the Gilmore Accounts, showing Banking Facilities

1862					
Oct.	3	H. Prentiss & Co.	Cr.		
		By note on 6 mos. from July 20/62			$665.35
	4	M. E. Reeves & Co. ch. from Raynham	$1,100.68		
		E. A. Goodnow " " "	262.29		
		M. E. Dittsman " " "	315.08		
	4	Wm. Claflin & Co.	Cr.		
		By check Reserved in store			85.00
	10	C. C. Fuller Reserved in Store	Cr.		
		By check	32.25		
		" Dis. 5 off	1.75	34.00	
	11	Potter, Hitchcock & Co.	Cr.		
		By check	109.44		
		" Dis. 5 off	5.76	115.20	
	14	M. E. Reeves Check from Raynham	1,260.		
Oct.	15	S. Wilder	Cr.		
		By cash for Bank Goods			42.00
	15	W. B. Seaver	Cr.		
		By cash for Bank Goods			42.00
	17	A. P. Hoover	Cr.		
		By Draft American Bk. U. S.	108.30		
		" Dis. 5 off	5.70	114.00	

APPENDICES 231

1862
Oct. 18 Jane & Slanusa
          Ch. from Raynham.......................... $1,343.06
      18 W. S. Lester   Cr.
          By J. H. Lester's Ch. for note................. $400.00
      18 T. W. Howard & Bro.   Cr.
          By Drafts on a/c.............................   200.00
      22 H. A. Ball
          By cash...................................... 22.80
          " Dis. 5 off.................................   1.20   24.00
      24 Philip Ford & Co.   Cr.
          By check.................................... 196.89
          "   " 5 off................................ 10.36   207.25
      29 Thos. Caden   Cr.
          By draft.................................... 287.52
          " Dis. 5 off................................ 15.13   302.65
      30 Potter, Hitchcock & Co.   Cr.
          By cash..................................... 175.56
          " Dis. 5 off................................   9.24   184.80
      31 E. A. Hendry Draft from Raynham............. 1,102.08
Nov.   1 Wm. North   Cr.
          By draft on a/c..............................   800.00
       3 Robert Morris & Co.   Cr.
          By Note at 8 mos. from Oct. 25/62............ 1,140.00
          "   "   "   "   " Nov. 3/62.............   960.00 2,100.00
       3 Nash & Fogg   Cr.
          By note at 8 mos. from Oct. 10/62............   787.95
          "   "   " 9 "   "   "   12/62............   790.00 1,577.95
       6 John B. Alley & Co.   Cr.
          By ch..   1,500.00

XXIV

EXCERPTS FROM THE KIMBALL AND ROBINSON PAPERS,[1] BROOKFIELD

Saturday, December 11, 1858

W. H. White,   Cr.
    By 60 pr. S. bottd............................. $8.50
    " 120 " " Bound..............................   3.10
                                                Dr.
    To cash...   31.99
Wm. Robinson   Dr.
    To 1 Lot Figans.................................   .42
Nelson L. Elmer   Dr.
    To 60 pr. ½ D. S. Calf Boots...............13/2 @ 70 42.00
    " 60 " 5/10 Childs Calf Boots................ @ 40   24.00
                                                                 $66.00

[1] These account books are owned by Mr. Henry E. Twitchell of North Brookfield.

Saturday, December 11, 1858

Bigelow & Hoagland	Dr.	
To 60 pr. 3/7 Spry 1 Calf 2.82½		$49.50
Cyrus Webber	Cr.	
By 60 pr. S. Bound		1.50
	Dr.	
To cash		2.26
A. Ainsworth	Dr.	
To cash (Del Webber)		.67
Dexter Henshaw	Cr.	
By 60 pr. S. Bound		1.20
	Dr.	
To thread		.06
Hiram O. Newton	Dr.	
To cash (Del Shaw)		23.00
Dexter Rice	Cr.	
By 60 pr. S. Bound		1.50
To cash	Dr.	4.39
G. W. Prince	Cr.	
By 180 pr. S. Bottd		21.00
	Dr.	
To cash		25.00
Ainsworth & Hawes[1]	Cr.	
By 60 pr. S. Bottd (Walker)		6.00
Wm. S. Pike	Cr.	
By 60 pr. S. Bound		1.50
	Dr.	
To cash		2.06
Henry H. Adams	Dr.	
To thread		.05
Wm. T. Lamb	Cr.	
By 120 W. S. Bottd		17.00
By 180 " S. Bound		4.80 21.80
	Dr.	
To cash		23.88
W. T. Lamb	Dr.	
To cash		.12
Mary Flagg	Cr.	
By 60 pr. S. Bound		1.20
	Dr.	
To cash		1.20

[1] One wonders if this was the firm name of a "bottoming gang." Walker is evidently the name of the man who collected the money.

APPENDICES 233

Saturday, December 11, 1858

Charlotte Howe	Dr.	
To cash (on a. c. H. Howe)		$5.35
J. B. Bellows	Dr.	
To order (A. H. Hawes)		5.00
Ainsworth & Hawes	Cr.	
By J. B. Bellows		5.00
Cash pd. for Can ? .75 Book 1.00		1.75
Augusta Walker	Cr.	
By 60 pr. S. Bound		1.60
G. W. Prince	Dr.	
to 5 lasts @ 15		.75

June 27, 1860.

Mary Ann Belcher	Cr.		
By 60 pr. S. Bound		1.30	
Geo. W. Wellman	Cr.		
By 120 pr. S. Botd		16.00	
	Dr.		
To cash		13.00	
Cyrus Webber	Dr.		
To cash		5.00	
T. Barton	Dr.		
To 600 pr. 3/7 L. L. Split (c) 70		420.00	
To 300 pr. 3/7 O. L. "		195.00	615.00
Chancy Whittemore	Cr.		
By 60 pr. S. B. & Botd[1]		10.10	
	Dr.		
To cash		9.99	
Oliver P. Kendrick	Dr.		
To cash		27.00	
A. Hubbard	Cr.		
By 180 pr. S. Bound		2.80	
	Dr.		
To cash		4.30	
Wm. B. Cooley	Cr.		
By 60 pr. S. Botd		8.00	
	Dr.		
To cash		5.00	
Franklin Sibley	Dr.		
To cash		25.00	
Henry Daniels	Cr.		
By 60 pr. S. Botd (poor)		— —[2]	

---

[1] Shoes bound and bottomed is meaning of S.B. & Botd.

[2] No amount given — probably no pay; just the comment "poor" is set down.

APPENDICES

June 27, 1860

P. S. Walker	Dr.	
To thread		$.19
David Johnson	Cr.	
By 60 pr. S. Bound		1.50
Braman Ward	Cr.	
By 60 pr. S. Bound		.90[1]
Samuel D. Bowen	Cr.	
By 60 pr. S. bound		.90
	Dr.	
To thread		.14
W. H. White	Cr.	
By 120 pr. S. Botd		16.00
Thomas Mambee	Dr.	
To cash		.90
	Cr.	
By 60 pr. S. Bound		.90
Oliver P. Kendrick	Cr.	
By 120 pr. S. Botd		12.00[2]
Oliver P. Kendrick	Dr.	
To 6 lasts		.84
G. A. Mention	Dr.	
To 2 lasts		.45
Ruben Slayton	Cr.	
by 120 pr. S. Bound		3.10
	Dr.	
To thread		.12
Unice Harrington	Cr.	
By 60 pr. S. Bound		1.50
Gardner Tyler	Cr.	
By 60 pr. S. Botd		8.50
	Dr.	
To cash		8.50
W. S. Cooper	Dr.	
To cash		10.00
Adrian Avery	Cr.		
By 1693½ G. D. Sole Lea. .... @20½		348.59
Cartage		.39 348.98

[1] No reason appears for different prices for same number of shoes bound.
[2] Note different price for bottoming — $16 for 120 pr., $12 for 120 pr.

APPENDICES

December 14, 1858.

George Millard   Dr.

To 60 pr. Mens calf 3/7	@87½	$52.50	
" " " " Kip 3/7	80	48.00	
" " " " Split 3/7	72½	43.50	
" 120 " Childs Calf 5/10	40	48.00	$192.00

December 18, 1858

Meade & Stowell   Dr.

To 180 pr. 3/7 Split	@70	126.00	
" 120 " 11/2 "	@55	66.00	192.00

January 1, 1859

Bigelow & Hoagland   Dr.

To 360 pr. L. L. Cf. Boots 3/7	@95	205.00	
" 360 " C. L. Cf " 3/7	@92½	333.00	
" 48 " split " Mc 3/7	@70	336.00	
120 " " " Spg 3/7	@67½	81.50	955.50

January 17, 1859

George Millard   Dr.

To 180 pr. Mns Calf 3/7	92½	166.50	
" 60 " " Kip 3/7	75	45.00	
" 120 " Childs Cf 5/10	40	48.00	259.50

January 22, 1859 [1]

Bigelow & Hoagland   Dr.

To 60 pr. 5/8 C. L. Cf boots C.	95½	51.00	
60 " 3/8 L. L. Cf " C	97½	58.50	
240 " 3/7 " " " C	95	228.00	
60 " 3/7 Col. " " C	92½	55.50	
60 " 3/7 " " " Spds	85	51.00	444.00

January 22, 1859

W. W. Griffith & Co.   . Dr.

To 120 pr. C. L. Cf. Boots 3/7	92½	111.00	
60 " Kip " 2d 3/7	@75	45.00	
120 " Cf. " Child. 5/10	42½	51.00	
160 " " " Spgs 5/14	@42	48.00	
60 " " " " 5/10	37½	22.50	277.50

On January 31, there is charged up "to Boston" a bill of $793 worth of boots, some with heels, others spgs (*i.e.*, Springheels?). Evidently there was a Boston office or retail store of this firm.

[1] On January 29, this same firm ordered $876 worth of boots.

XXV

EXCERPTS FROM THE TWITCHELL PAPERS [1]

Date 1861 when taken	Binder's name	No. of case	Kinds				Quality		Wom.Miss.Chil.
			Clf.	Kip.	Buff.	Split.	First	Second	
May 7	H. Sibley	500	"						"
Apr. 24	Wm. Maynard	501	"						"
" 25	G. N. Barnes	502	"						"
" 24	M. W. Cowley	503	"				L. L.		"
" 29	H. Sibley	504	"				L. L.	"	
May 3	E. Davis	505			"		"		"
" 4	D. Henshaw	506			"		"		"
" 4	W. W. Bowen	507			"		"		"
" 6	P. S. Walker	508			"		"		"
" 6	W. Palmer	509			"		"		"
May 4	D. Henshaw	510			"		"		"
" 6	Chas Hamilton	511	"				"		"
" "	Geo. R. Newton	512	"				"		"
" 10	P. Lackey	513	"				"		"
May 10	Geo. R. Newton	514	"				"		"
" 10	C. A. Chaffee	515	"				"		"
" 9	A. Nichols	516	"				"		"
" 9	M. A. Belcher	517	"				"		"
" 8	Wm. Nichols	518	"				"		"
" 7	E. A. Gay	519	"				"		"
" 10	C. Webber	520			"		"		"
" 10	D. Henshaw	521			"		"		"
" 15	J. P. Stearns	522			"		"		"
May 10	S. Campbell	523			"		"		"
" 10	E. Davis	524			"		"		"
" 13	H. Sibley	525			"		"		"
" 17	D. Coombs	526			"		"		"
" 17	P. S. Walker	527			"		"		"
" 17	P. Laskey	528							
" 17	D. Johnson	529							
" 17	E. Davis	530							
" 18	D. Henshaw	531							
" 21	H. Sibley	532							
" 22	Wm. Nichols	533							
" 22	Wm. Maynard	534							
" 21	M. A. Belcher	535							
" 21	C. Webber	536							
" 18	D. Henshaw	537							
" 15	F. Shaw	538							
" 13	B. Ward	539							
" 14	L. Russell	540							
" 15	F. Barnes	541							
" 17	M. Covley	542							
" 14	P. S. Walker	543							
" 15	E. A. Gay	544							

[1] Twitchell bought out Kimball & Robinson in 1861. These pages show advanced bookkeeping methods.

XXV

EXCERPTS FROM THE TWITCHELL PAPERS

When returned	Price paid	Date 1861 when taken	Bottomer's name	Kind	When returned	Price paid	Sold	Rmrks
May 13	1.20	June 5	Wm. B. Cooley		June 29	–		
" 28	1.20	Apr. 29	G. W. Barnes		May 28	–		
" 13	1.20	Apr. 29	G. W. Barnes		" 13	6.00		
" 7	1.30	" 24	W. B. Cooley	C	" 7	5.00		
May 4	1.30	May 7	W. B. Cooley		June 5	6.00		
May 17	1.60	May 4	G. P. Kendrick		May 16	–[1]		
" 10	1.50	" 15	F. Shaw		June 6	–		
" 18	1.60	Oct. 26	N. H. Delane		Nov. 8	–		
" 9	1.60	May 10	N. H. Delane		May 24	–		
June 6	1.60	" 20	Wm Palmer		June 6	–		
May 10	1.60	May 14	E. E. Chapin		May 25	–		
" 20	1.60	" 6	Ch. Hamilton		May 20	–		
June 7	1.60	" 6	Geo. R. Newton		June 7	–		
May 17	1.60	Sept. 18	Ch. Hamilton	T. S.	Oct. 12	–		
June 14	1.60	May 10	Geo. R. Newton		June 14	–		
May 18	1.60	" 20	H. May		May 24	–		
" 23	1.60	" 25	E. E. Chapin		June 1	–		
May 21	1.60	" 24	H. May		" 1	–		
June 8	1.60	" 22	A. Bemis		" 8	–		
May 17	1.60	" 25	F. Sibley		" 13	–		
May 21	1.60	" 24	H. May		" 1	–		
" 18	1.60	Oct. 23	J. Mitchell	½?	Nov. 7	–		
" 28	1.60	Sept. 14	N. H. Delane		Sept. 2	–		
" 24	1.60	May 24	J. F. Bemis		June 29	–		
Sept. 5	1.60	Sept. 6	N. H. Delane		Sept. 14	–		
May 21	1.60	May 22	J. W. Fittz		June 1	–		
" 24	1.60	May 17	Benj. Davis		May 24	–		
" 25	1.60	Oct. 29	H. May	½W	Nov. 7	–		
		Oct. 26	E. E. Chapin		" 4	–		
		Oct. 31	D. Shaw	½W	" 18	–		
		Sept. 26	Geo. Lavelley		Oct. 9	–		
		" 28	N. H. Delane		" 17	–		
		June 1	E. E. Chapin		June 12	–		
		Aug. 31	J. F. Bemis		Sept. 28	–		
		Oct. 4	E. E. Chapin		Oct. 14	–		
		June 1	J. W. Fitts		June 25	–		
		Oct. 20	H. May		Nov. 7	–		
		May 25	N. H. Delane		June 1	–		
		June 7	F. Shaw		Aug. 21	–		
		Oct. 30	Geo. R. Newton		Nov. 19	–		
		Oct. 1	F. Barnes		Oct. 14	–		
		June 7	F. Barnes		Aug. 15	–		
		Sept. 23	H. May		Oct. 1	–		
		Oct. 18	E. E. Chapin		Oct. 26	–		
		June 1	H. May		June 12	–		

[1] This – is meant for a ditto in the ledger and evidently stands for $6.00, the regular price.

Excerpts from the Twitchell Papers

Date 1864 when taken	Binder's name [1]	No. of case	Kinds Clf. Kip. Buff. Split.	Quality First Second	Wom.Miss.Chil.
		2	Cases Boys Split	1 to 5	
		1	" Youths "	9 — 13	
		3	" Boys "	1 — 5	
		4	" " "	1 — 15	
		1	" " 2d	1 — 5	
		1	" Youths 2d "	9 — 13	
		2	" Boys 1st "	1 — 5	
		2	" " 2d "	1 — 5	
		4	" " 1st"	1 — 5	

Volume and Distribution of Business, and Methods of Transportation of Henry E. Twitchell

From Order Book, February 23, 1861 to Nov. 28, 1865.

Firm	Dif. sizes kinds & prices amount	1861 date	Shipped via
Wadsworth & Wells Chicago	69 cases	Feb. 23/61	Merchants Despatch
Henderson & Co., C. M. Chicago	3 "	"	" "
Fargo & Bill, Chicago	3 "	"	L. S. & M. C. R. R.
V. Barber, Decatur, Ill.	3 "	Mar. 5/61	Great West. Despatch
O. Rugg, Bloomington, Ill.	2 "	"	" " "
C. M. Lee & Bro. Boston	24 "	Apr. 10– Oct. 30/61	?
W. W. Griffin & Co., Toledo, Ohio	6 "	Mar. 8/61	Great West. Despatch
Bassett & Emmal, Lexington, Ky.	5 "	Mar. 4/61	Merchants Despatch care of Mr. Burney Cincinnati, Ohio
George Millard, North Adams, Mass. Ranged from 30c. to 95c. a pair	22 "	Mar. 28 & Aug. 17/61	
Geo. O. Catlin, Leavenworth, Kan.	22 cases	Aug. 19 & Nov. 5/61	Great West. & Mich. Cent. line Send R. R. (?) to Otis Kimball, 21 State St. for Bill Lading
C. M. Henderson & Co. Chicago	34 cases	Aug. 15– Oct. 28, 1861	L. S. & M. C.

[1] These shoes were no longer put out to bind, only to bottom. Probably they were stitched in the factory. The same ledger was used, with its old forms, even after a new custom arose.

APPENDICES

Excerpts from the Twitchell Papers

When returned	Price paid	Date 1864 when taken	Bottomer's name	Kind	When returned	Price paid	Sold	Rmrks.
		Apr. 1	G. R. Newton		Apr. 12	–		
		" 6	Austin Nichols		" 14	–		
		" 6	Geo. H. Levalley		" 16	–		
		" 12	Geo. R Newton		" 19	–		
		" 12	" " "		" "	–		
		" 15	A. Nichols					
		" 14	B. F. Chever		" 30	–		
		" "	" "		" 30	–		
		" 15	Geo Levalley		June 13			

From Order Book, February 23, 1861 to Nov. 28, 1865.

Firm	Dif. sizes kinds & prices amount	1861 date	Shipped via
John Powell, Dayton, Ohio	15 cases	Aug. 17– Nov. 6 1861 on 6 mos. time	Merchants Despatch
Bassett & Emmal, Lexington, Ky.	13 cases from 30 to 70c.	Aug. 17/61 on 6 mos. time c/o	Merchants Despatch
Baldwin Studwell & Fi N. Y.	2 cases at 90c.	Aug. 17/61 on 6 mos. time	
D. Rugg & Noyes, Champaign, Ill.	8 cases	Aug. 20 to Nov. 15/61	Great West. Despatch
A. B. Smith & Co., Madison, Ind.	21 cases	Aug. 27– Dec. 5, 1865	" via Indianapolis
Miller & Parsons, Cleveland, Ohio	9 cases { 4 cases 51–60 from 55– 90c. }	Sept. 9 & 16 /61 time cash note	Merchants Despatch
Marcy & Haynes, Hartford, Ct.	15 cases from 65 to 92	Sept. 13, Nov. 7 & Dec. 5/61	
Howes, Hyatt & Co., N. Y.	3 cases	Sept. 18/61	
A. & F. Reed, N. Y.	4 " " " 65c. to 82c.		
N. L. Emer, Greenfield, Mass.	2 cases	Sept. 19/61	
O. Rugg, Bloomington, Ill.	5 cases	Sept. 24 Oct. 17/61	Grt. West. Despatch
Rufus Elmer, Springfield	15 cases	Sept. 20 & Oct. 7/61 on 6 mos. time	

APPENDICES

From Order Book, February 23, 1861 to Nov. 28, 1865.

Firm	Dif. sizes kinds & prices amount	1861 date	Shipped via
Mead, Stowell & Co., N. York from 67 to 95c. per pair	31 cases	Nov. 26 & Dec. 23/61 on 6 mos. time	via New Haven
Crowell & Childs, Cleveland, Ohio	3 cases	Nov. 2/61	Merchants Despatch
Atkins, Strak & White Milwaukee	3 cases	Nov. 27/61 on 6 mos. time line	G. W. & D. & M.
V. Barker, Decatur, Ill.	3 cases	Nov. 8 & Dec. 12/61	G. West. Despatch
Rugg & Noyes, Champaign, Ill.	5 cases	Jan. 10	Grt. West. Despatch
Geo. O. Catlin, Leavenworth, Ky	24 cases	Jan. 10–Mar. 28	G. W. & M. C. line
John Powell, Dayton, Ohio	26 cases	Feb. 13	Merchants Despatch
M. P. Lancaster, Lexington, Ky	2 cases send bill to American House this week	Feb. 13	Merchants Despatch c/o E. Taylor Cincinnati, Ohio
Hewett, Burgest & Co., Cleveland, Ohio	13 cases	Feb. 13 & Apr. 2	Merchants Despatch
R. M. Pomeroy & Co., Cincinnati, Ohio	6 cases "Mark cases on back end & not strap." Nett prices. Send R. R. receipt	Feb. 13	" "
Fargo & Bill, Chicago, Ill.	36 cases Send R. R. Receipt to F. G. Faxon Boston for bill lading	July 14	L. S. & M. S. R. R. Pay frt. to Worcester & ch. F. & B. Ship to Wor. to be delivered to F. & W. R. R.
Hills & Goodman, Hartford, Co.	2 cases	Feb. 13	
Wells & Christie, N. Y.	150 cases	Mar. 3 & 8 Aug. 20	
R. & J. Cummings, Toledo, Ohio	11 cases	Feb. 24 & July 14	Merchants Desp.
V. Barber, Decatur, Ill	12 cases	Mar. 11	G. W. Despatch.
A. B. Smith & Co., Madison, Indiana	11 cases	Mar. 4 & Oct. 11	Merchants Desp. via. Indianapolis
C. M. Lee & Bros. Boston	38 cases 40 cases	Apr. 2 & Aug. 18 Nov. 5	
F. C. & D. Wells, Chicago, Ill.	May 2, July 14 Nov. 4 71 cases		Merchants Desp & some by L. S. & M. S. R. R. Gov. tax added

APPENDICES

From Order Book, February 23, 1861 to Nov. 28, 1865.

Firm	Dif. sizes kinds & prices amount	1862–3 date	Shipped via
C. M. Henderson & Co., Chicago, Ill.	46 cases	May 7 July 11, 17 Sept. 13	L. S. & M. S. R. R. Ship to Worces. & some Merch. Desp.
Marcy & Haynes, Hartford, Ct.	44 cases	July 3	
Rugg & Noyes, Champaign, Ill.	35 cases Ship	July 5 Aug. 15	G. W. Despatch
John Powell, Dayton, Ohio	25 cases	July 7 & Nov. 5	Merchants Despatch
Mead, Stowell & Co., N. Y.	45 cases	July 17 & Oct. 28	via Pate & N. L. W. R. R.
Geo. O. Catlin, Leavenworth, Kan.	60 cases	July 19	G. W. & M. C. Line.
O. Bailey, White Cloud, Ka.	9 cases Send R. R. receipt to Peter McIntyre, & Co. Boston	July 26	Grt. West. Despatch via Worces. & Prov. R. R.
Huett, Burgest & Co., Cleveland, Ohio	9 cases	July	Merchants Desp.
V. Barber, Decatur, Ill.	143	Mar. 14, 1862	" "
W. W. Cane, Greenfield, Mass.	3 cases	?	
O. Powers, Decatur, Ill.	4 cases	Nov. 11/62	G. W. Desp.
Burgest & Adams, Cleveland, Ohio	9 cases	Feb. 1863	Merch. Despatch
Atkins, Steale & White Milwaukee	23 cases price ranges from $1.15 to 87c.	Sept. 25, 1863	G. W. & D. M. Line
O. Rugg, Bloomington, Ill.	1 case	Feb. 1863	G. W. Despt.
D. Rugg, Champaign, Ill.	12 cases (60 pr. cases)	? 1863	L. S. & M. S. via Chicago Ship to Worcester
Nickerson, Harris & Moseley Philadelphia	16 cases range $.95–$1.15	July 23 & Oct. 14, 1863	via N. Y. & P. R. R.
Conant, Warren & Co. Boston	7 cases	Aug. 10, 1863	
C. M. Lee & Bros. Boston	13 cases	Aug. & Sept. 1863	
Messrs F. C. & M. D. Wells	40 cases Some as high as $1.40 pair Goat	Nov. 3, 1863	Union Desp.
Mead, Stowell & Co., N. Y.	20 cases Some as high as $1.30	Sept 25 & Dec. 11, 1863	

APPENDICES

From Order Book, February 23, 1861 to Nov. 28, 1865.

Firm	Dif. sizes kinds & prices amount	1863-4 date	Shipped via
R. Elmer, Springfield, Mass.	1 case 1 case 3/7 L. L. Calf. 1.15	Sept. 26	
Baldwin, Fisher & Co., N. Y.	21 cases	Oct. 3, 9 & Nov. 17, 1863	
Claflin, Mellen & Co., N. Y.	5 cases	Oct. 3, 1863	
Geo. O. Catlin, Leavenworth, Kan.	32 cases	July 11, 1863	G. W. & M. C. Line
C. M. Henderson & Co., Chicago	29 cases prices 70c., 87c., $1.05, $1.10	Jan. 29, 1863	Mer. Despatch
F. C. & M. D. Wells, Chicago	56 cases L. L. Buff	Feb 3 & Apr. 9, 1863	L. S. & M. S. R. R.
Wells & Christie, N. Y.	15 cases	Oct. 3/63	Nett 30 days
John Powell, Dayton, Ohio	8 cases	Nov. 9/63	Mer. Desp.
Messrs. Hoagland, Dubois & Magovern, N. Y.	57 cases	Dec. 19, 1863	
Wells & Christie, N. Y.	15 cases	Mar. 3 & 16	
John Powell, Dayton, Ohio	12 cases 2 cases	Jan. 3, 1864 Jan. 29	Mer. Desp.
Hoagland, Dubois & Magovern, N. Y.	32 cases	Mar. 16	
Howes, Hyatt & Co., N. Y.	4 cases	Jan. 1	
R. & J. Cummings, Toledo, Ohio	14 cases	July 29	?
Sterling & Franks, Phil	4 cases	Feb.	P. & N. L. R. R. via C. & A. R. R.
Bailey & Noyes, White Cloud, Ks.	7 cases some as high as $1.85	Feb. 1 July 13, & Oct. 24	Union Despatch
Gasten & Stowell, N. Y.	71 cases 75 " (These were mostly Youths and Boys)	June 10 Aug. 12	
Mead & Stowell, N. Y.	64 cases	June 15 & 20 Aug. 12 & Nov. 17	
Messrs. A. & A. G. Trask, N. Y.	127 cases	June 28 Mar.	
Hedges & Powers, N. Y.	47 cases ranging $1.05 to $1.87 cg. 10 c 3/7 L. L. Buff Lace	Nov. 18, 28 & 30	

APPENDICES

From Order Book, February 23, 1861 to Nov. 28, 1865.

Firm	Dif. sizes kinds & prices amount	1864–5 date	Shipped via
E. B. Fuller, Fredonia, N. Y.	3 cases	July	Send to Dunkirk via E. Albany
D. & W. B. Bailey, Champaign, Ill.	5 cases	July 20, 1864	C/o P. McIntyre, Boston.
G. O. Catlin, Leavenworth, Ks.	8 cases	July 25	Ship to H. A. Ball Boston.
J. S. Christie, N. Y.	62 cases	Aug. 12	
D. Rugg, Champaign, Ill.	6 cases of 60 pr. each 2 cases	Aug. 15	Merchants Desp.
V. Barber, Decatur, Ill.	13 cases	Aug. 18	" "
Bill Wheelock & Co., N. Y.	9 cases	Oct. 21	via N. L. W. & P. R. R.
Mabie, Manly, Murray & Morgan, N. Y.	26 cases	Nov. 18	
James French, N. Y.	36 cases	Nov. 18 & 28	
W. A. Ransome & Co., N. Y.	2 cases	Nov. 18	
Washington Olds Agt, Albany, Ill. c/o Snider & Co., Fuller, Ill.	16 cases	Aug. 23	Merchants Desp.
Messrs Porter & Higby, N. Y.	19 cases i. e. 1140 pair	Mar.	
L. E. Schoonmaker & Co., N. Y.	2 cases		
R. & J. Cummings, Toledo, Ohio	6 cases	Feb. 1	Merchants Desp. Steamer on Lake
Mark N. T. Co., c/o J. E. Bacon, Worcester	55 cases	July 14 & Nov. 10	
G. O. Catlin, Leavenworth, Ks.	27 cases	July 11, 1865	Send to A. Strong & Co.
J. S. Christie, N. Y.	26 cases	July 14 Sept. 2	
D. Rugg, Champaign, Ill.	29 cases	July 22 & Aug. 1	Merchants Desp.
V. Barber, Decatur, Ill.	33 cases Boys, Childrens, Youths, men — all dif. kinds.	Aug. 21	Merchants Desp.
Bill Wheelock & Co., N. Y.	20 cases	Nov. 27	via N. L. W. & P. R. R.

APPENDICES

From Order Book, February 23, 1861 to Nov. 28, 1865.

Firm	Dif. sizes kinds & prices amount	1865 date	Shipped via
Mabie, Manly, Murray & Morgan, N. Y.	127 cases	July 17– Aug. 16, 1865	
John Powell & Co., Dayton Ohio	11 cases range .65 to $1.90	Feb. 2, 1865	via Worcester & Providence
Atkins, Steale & White, Milwaukee	65 cases boys, youths & children	July 18, 1865	N. T. Co. via W. & N. R. R. J. E. Bacon Agt
C. M. Lee & Bros., Boston	30 cases	July 27 & Aug. 23	
Messrs. A. B. Smith & Co. Madison, Indiana	15 cases	July 26	G. W. Despatch via Cincinnati Care of Mailboat
O. Rugg, Bloomington, Ill.	46 cases	Sept. 13	Merchants Despatch
Mead, Townsend & Andrews, N. Y.	161 cases	Aug. 2 & 23 & Oct. 18	
Howes, Hyatt & Co., N. Y.	202 cases mostly boys & youths	Aug. 1865	
Marcy & Haynes, N. Y.	18 cases	Aug. 19, 1865	
Burgest & Adams, Cleveland	20 cases "Take all I have of Buff & Split Bal."	Aug. 26 & Sept. 2	Merchants Despatch
Hoagland, Dubois & McGovern, N. Y.	73 cases	Sept. 14 & Oct. 17, 1865	
French, Powell & Co., N. Y.	31 cases	Sept. 17, 1865	
Cutter, McIntosh & Co., Springfield	42 cases	Oct. 19 Nov. 13 Dec. 6	
	Youths & Boys		
Messrs. Fargo, Bill & Co., Chicago, Ill.	43 cases Boys and Youths	Nov. 8, 1865	L. S. & M. S. Red Line
Chas. F. Parker & Co., Boston	10 cases	Nov. 15	
Baldwin, Fisher & Co., N. Y.	4 cases	Nov. 27	
Smith & Maynard, N. Y.	4 cases	Nov. 28	

XXVI

The McKay Machine

Mr. Lyman R. Blake made a sworn statement in 1872 concerning his invention of the so-called McKay machine in connection with his application for the extension of his patent of July 6, 1858, as follows: [1]

I entered the firm of Gurney & Mears, in South Abington, in 1857. At that time all boots and shoes made where I lived had their soles united with the uppers by pegging or nailing. It was not until some time after I made my invention that I ever saw a shoe sewed by hand. To increase the business of my firm (Gurney, Mears & Blake), it was desirable to manufacture sewed shoes, and as there were no workmen in and about Abington skilled in such work, I began to consider whether or not I could devise machinery for sewing soles to boots and shoes. When I had clearly in mind a conception of my invention, which was afterwards patented by me, I went to my partners for consent to build a machine. At first they objected, but afterwards consented that I might build it with my own money, provided the firm should have the use of the machine.

The machine was built — how, when and where I shall prove hereafter — and was used by and for the firm — to what extent and with what success, will also be proved. . . . After my return to Massachusetts,[2] being in improved health and desirous of aiding in the introduction of my sole sewing invention to public use, I joined Mr. McKay in 1861, and with some exceptions, which I will state, have kept such connections with him. In said time I have labored in every way possible to increase the capacity for work of machines embodying my invention, to instruct operators to run them, to induce manufacturers to use them and to remove from the public mind all prejudice concerning the machines. I have also made many inventions relating to the business of sewing soles by mechanism. All of which have, by deed of assignment before referred to, passed to the ownership of Mr. McKay and his successors.

There are in the United States sixteen of such patents for my individual invention, and six in each of which I am one of joint inventors. Some of these relate to turned soles. I have also been employed for about a year on an invention for nailing soles, which invention has passed to McKay and others.

After my return from Staunton, Va., my first service rendered in connection with my invention was in experimenting with the machines built by Mr. McKay, and afterwards in working them in sewing the shoes for the Mass. Light Artillery Battery. Next I sewed the shoes for one or two regiments, and afterwards many thousands of pairs of army shoes; all the time

[1] From a copy of the report before Patents Case No. 20,775.

[2] He had been in Staunton, Va., seeking to regain his health.

keeping watch of the operation and wear of the machines, and advising Mr. McKay wherein they were deficient and how they should be improved.

This army work was very heavy, and was undertaken with a view to test the machines, that they might be rendered nearly perfect before they were sold to the public.

The patents before alluded to have been taken between the date of my first patent and the present time, and include improvements in and additions to the sewing machine patented in 1858; machines for waxing thread and for channeling soles preparatory to sewing; devices for lasting; for sewing; improvements in the arrangements of the soles and vamps; to adapt them for sewing by mechanism, etc.

I have traveled largely over the United States, and have instructed men how to use the sewing machines and their adjuncts, and have organized the forces in the workshops, so that in adopting the new system confusion and delay have been prevented. I have endeavored to extend the knowledge of my invention, and have aided manufacturers in every way in my power to do a safe and profitable business with my invention, and have instructed them and their workmen how to use it so as to produce work of high quality. Prior to my invention, no machine existed for sewing soles to boots and shoes.

XXVII

Excerpts from the Batcheller Accounts, Brookfield

Accounts for a Single Year 1870

Dr.				Boston Store [1]			Cr.
1870				1870			
Jan. 31	To amt.			$138.00	Jan. 6	By Stock a/c	$261,328.52
Mar. 31	"	"		35.00	Jan. 31	" Amt.	28,457.21
May 31	"	"		102.60	Feb. 28	" "	54,571.05
Oct. 31	"	"		260.00	Mar. 31	" "	45,399.31
Nov. 30	"	"		67.27	Apr. 30	" "	32,444.24
Dec. 8	"	"		154.35	May 31	" "	41,800.87
"	"	" sds.		62.76	June 30	" "	35,963.85
"	"	" "		29,256.30	July 30	" "	38,343.41
"	"	" stock act		1,700,342.41	Aug. 31	" "	58,324.55
"	"	" "		7,938.70	Sept. 30	" "	66,380.96
"	"	" bal.		268,956.97	Oct. 31	" "	65,226.75
					Nov. 30	" "	58,797.48
					Dec. 8	" "	32,773.37
					"	"	1,187,502.79
				$2,007,314.36			$ 2,007,314.36

[1] The firm's store in Boston, where the factory product was disposed of.

APPENDICES

Cash

Dr.					Cr.	
1870				1870		
Jan.	7	To bal.	$21,460.68	Jan. 31	By amt.	$37,925.16
"	31	" amt.	30,099.70	Feb. 28	" "	53,794.55
Feb.	28	" "	55,726.93	Mar. 31	" "	52,950.07
Mar.	31	" "	42,073.22	Apr. 30	" "	33,523.65
Apr.	30	" "	35,539.19	May 31	" "	34,318.06
May	31	" "	42,469.20	June 30	" "	39,116.36
June	30	" "	37,584.39	July 30	" "	48,892.12
July	30	" "	39,726.42	Aug. 31	" "	47,250.95
Aug.	31	" "	58,291.49	Sept. 30	" "	56,321.73
Sept.	30	" "	65,440.44	Oct. 31	" "	74,827.26
Oct.	31	" "	68,243.85	Nov. 30	" "	53,620.17
Nov.	30	" "	53,669.34	Dec. 8	" "	35,910.08
Dec.	8	" "	33,163.17		Bal.	15,037.86
			$583,488.02			$583,488.02

Cutting Dept.

Dr.				Cr.	
1870			70		
Jan. 31	To amt.	$5,240.95	Jan. 6		$43,412.43
Feb. 28	" "	5,673.41	included in bal. taken Jan. 6th		
Mar. 31	" "	3,807.35			
Apr. 30	" "	2,526.40			
May 31	" "	3,537.24	June 9	By amt.	$1.50
June 30	" "	4,326.80	July 5	" "	.62
July 30	" "	4,915.49	" 8	" "	.97
Aug. 31	" "	5,915.50	Aug. 13	" "	.50
Sept. 30	" "	5,714.64	" 17	" "	2.50
Oct. 31	" "	5,734.81	" 22	" "	2.75
Nov. 30	" "	403.67	Sept. 6	" "	1.00
Dec. 8	" "	6,454.34	Oct. 7	" "	.56
			Oct. 25	" "	.50
			Nov. 9	" "	.96
			" 12	" "	.50
			Dec. 7	" "	24.90
			" 8		54,213.34
		$54,250.60			$54,250.60

BOTTOMING DEPT.

	Dr.				Cr.
1870					
Jan. 31	To amt.	$28,031.02	Dec. 8	By amt.	$272,314.48
Feb. 28	" "	30,252.56			
Mar. 31	" "	21,144.06			
Apr. 30	" "	12,783.22			
May 31	" "	16,308.26			
June 30	" "	20,805.03			
July 30	" "	23,954.06			
Aug. 31	" "	28,467.00			
Sept. 30	" "	28,800.77			
Oct. 31	" "	29,095.06			
Nov. 30	" "	27,741.07			
Dec. 8	" "	4,932.37			
		$272,314.48			$272,314.48

STOCK ACCOUNT

	Dr.				Cr.
1870			1870		
Jan. 7	To Boston Store	$261,328.52	Jan. 8	By Amt.	$111.70
14	" Sds	479.70	" 11	" "	2.85
15	" Amt.	80.00	" 20	" "	6.54
28	" "	13.00	" 24	" "	3.66
31	" "	467.00	" 26	" "	.42
"	" "	549.68	" 31	" "	206.91
"	" "	3.50	Feb. 5	" "	149.35
"	" "	32.50	" 11	" "	5.27
"	" "	197.91	" 17	" "	311.40
Feb. 4	" "	49.62	" 21	" "	.80
11	" Sds.	294.15	" 28	" "	265.23
16	" "	67.20			
"	" Amt.	145.74			$1,064.13
21	" "	228.00			
28	" Sds.	69.20			
"	" Amt.	1,546.19			
"	"	49.00			
"	"	46.00			
		$265,646.91			
1870			1870		
Dec. 8	To amt.	$282,532.77	Dec. 8	By amt.	$5,521.20
" "	" "	40.42	" "	" "	57.84
" "	" Sds	517,101.43	" "	" "	1,700,342.41
		12,929.39			7,938.70
		1,179,662.69		Bal.	280,349.45
		1,942.90			
		$1,994,209.60			$1,994,209.60

APPENDICES

	Dr.	FREIGHT		Cr.
1870			1870	
Jan. 31	To amt.	$1,200.30	May 4 By amt.	$254.67
Feb. 28	" "	1,032.76	Sept. 8 " "	2.20
Mar. 31	" "	901.76	" " " "	2.45
Apr. 30	" "	703.89	Oct. 11 " "	9.00
May 31	" "	1,199.20	Nov. 8 " "	11.00
June 30	" "	1,096.33	Dec. 8 " "	14,536.56
July 30	" "	1,650.25		
Aug. 31	" "	1,310.97		
Sept. 30	" "	1,622.61		
Oct. 31	" "	1,402.14		
Nov. 30	" "	2,304.50		
Dec. 8	" "	391.17		
		$14,815.88		$14,815.88

	Dr.	RUNNING EXPENSES		Cr.
1870			1870	
Jan. 31	To amt.	$171.70	Apr. 8 By amt.	$5.00
Feb. 28	" "	157.65	" 12 " "	20.72
Mar. 31	" "	1,143.80	Dec. 8 " "	3.00
Apr. 30	" "	826.57	" " " "	5,485.94
May 31	" "	654.36		
June 30	" "	1,210.88		
July 30	" "	164.25		
Aug. 31	" "	67.50		
Sept. 30	" "	270.10		
Oct. 31	" "	447.21		
Nov. 30	" "	347.64		
Dec. 8	" "	53.00		
		$5,514.66		$5,514.66

	Dr.	GENERAL EXPENSES		Cr.
1870			1870	
Jan. 31	To amt.	$182.76	Sept. 21 By Amt.	$.40
Feb. 28	" "	369.77	Oct. 18 " "	40.00
Mar. 31	" "	304.06	Dec. 8 " "	21.00
Apr. 30	" "	258.98	" " " "	7,139.15
May 31	" "	299.20		
June 30	" "	276.98		
July 30	" "	341.26		
Aug. 31	" "	304.16		
Sept. 30	" "	446.23		
Oct. 31	" "	707.74		
Nov. 30	" "	2,227.04		
Dec. 8	" "	1,482.37		
		$7,200.55		$7,200.55

	Dr.		TEAMING			Cr.
1870				1870		
Jan. 31	To amt.		$418.00	May 4	By amt.	$164.55
Feb. 28	"	"	382.37	Dec. 8	" "	4,672.64
Mar. 31	"	"	322.59			
Apr. 30	"	"	316.23			
May 31	"	"	365.56			
June 30	"	"	311.26			
July 30	"	"	579.54			
Aug. 31	"	"	430.28			
Sept. 30	"	"	489.63			
Oct. 31	"	"	479.58			
Nov. 30	"	"	575.42			
Dec. 8	"	"	166.73			
			$4,837.19			$4,837.19

	Dr.		INTEREST			Cr.
1870				1870		
Jan. 31	To amt.		$.29	Oct. 3	By amt.	$3.22
Feb. 28	"	"	41.22	" 17	" "	.94
Mar.	"	"	43.74	" "	" "	4.66
Apr.	"	"	36.47	" 31	" "	4.37
May	"	"	21.24	Dec. 8	" "	195.10
June	"	"	39.92	" "		725.02
July	"	"	36.39			
Aug. 31	"	"	38.48			
Sept.	"	"	49.49			
Oct.	"	"	171.93			
Nov.	"	"	34.20			
Dec. 8	"	"	419.94			
			$933.31			$933.31

APPENDICES

Dr.				MACHINERY			Cr.	
1870				1870				
Jan.	7	To bal.	$27,420.42	Jan.	26	By amt.		$.75
"	31	" amt.	269.11	"	31	" "		2.00
Feb.	28	" "	308.71	Feb.	28	" "		2.00
Mar.		" "	604.72	Mar.	17	" "		3.40
Apr.		" "	281.28	"	29	" "		2.00
May		" "	237.76	May	7	" "		2.00
June		" "	208.61	"	31	" "		2.00
July	30	" "	159.70	June	3	" "		1.40
Aug.		" "	317.31	"	9	" "		9.72
Sept.		" "	445.00	"	30	" "		2.00
Oct.		" "	949.68	July	28	" "		1.40
Nov.		" "	604.26	"	30	" "		2.00
Dec.	8	" "	635.22	Aug.	23	" "		5.61
"	"	" "	4,208.46	"	31	" "		4.00
				"	"	" "		2.00
				Sept.	30	" "		2.00
				Oct.	12	" "		4.00
				"	18	" "		13.50
				"	"	" "		12.00
				"	"	" "		76.00
				"	22	" "		6.00
				"	31	" "		2.00
				Nov.	30	" "		2.00
				Dec.	8	" "		27.80
				"	"	" "		12,929.39
						" bal		23,533.27
			$36,650.24					$36,650.24

Dr.				INSURANCE			Cr.	
1870				1870				
Feb.	11	To cash	$20.00	Dec.	8	By Boston Store		$447.75
Mar.	12	" "	409.25					
"	31	" "	8.50					
Oct.	14	" "	10.00					
			$447.75					$447.75

Dr.				GENERAL WORK			Cr.	
1870				1870				
Jan.	29	To amt.	$.50	Feb.	14	By amt.		$1.87
"	31	" "	20.37	Aug.	19	" "		100.00
"	"	" "	9,994.57					5.00
Feb.	5	" "	15.75					121.13
		and so on to a total of	129,852.53			By bal.		129,624.53
Dec.	8	To bal.	$129,624.53					$129,852.53

APPENDICES

Dr.	Box Account	Cr.
1870	1870	
Jan. 31	Dec. 8 By amt.	$20,261.48
to		
Dec. 8 total $20,261.48		$20,261.48

Dr.	Real Estate	Cr.
1870		
Feb. 28 To amt.	$209.07	1870
Mar. 31 " "	390.77	May 26 By amt. $26.60
	154.11	Aug. 31 " " 302.16
	643.12	Sep. 7 " " 8.20
		Oct. 3 " Sds 1,095.85
		" 5 " amt. 3.00
	3,806.62	" 18 " " 10.00
	56.19	Nov. 15 " " 47.70
	99.16	" 30 " " 1,743.13
	190.70	Dec. 8 " " 345.00
Oct. 31	7,532.18	" " " " 40.03
Dec. 8	12,213.44	" " " " 8.00
		21,665.69
total	$25,295.36	$25,295.36

Dr.	Real Estate Expenses	Cr.
1870		1870
Jan. 25 To cash	$15.75	Dec. 8 By amt. $1,116.97
and so on in detail		
to a total		
Dec. 8	$1,116.97	$1,116.97

	Dr.		House Rent				Cr.
1870				1870			
Apr. 28	To amt.		$1.87	Jan. 29	By amt.		$275.00
				Feb. 25	" "		265.20
	bal.		3,631.64	Mar. 31	" "		263.57
				Apr. 8	" "		54.25
				" 11	" "		3.00
				" 29	" "		191.50
				" 30	" "		24.00
				May 28	" "		203.72
				June 27	" "		224.74
				July 30	" "		226.84
				Aug. 30	" "		229.25
				Sept. 23	" "		229.25
				Oct. 27	" "		222.75
				Nov. 25	" "		199.25
				Dec. 7	" "		958.33
				" 8	" "		62.86
			$3,633.51				$3,633.51

	Dr.		Charles H. Stoddard				Cr.
1871				1871			
June 1	To cash		$60.00[1]	May 31	By work		$60.90
" 13	" "		20.00	June 30	" "		50.70
July 1	" "		30.00	July 31	" "		43.80
" 31	" "		43.00	Aug. 31	" "		78.90
Sept. 1	" "		80.00	Sept. 30	" "		64.65
" 28	" "		65.00	Oct. 31	" "		45.00
Nov. 1	" "		45.00	Nov. 29	" "		44.10
" 29	" "		45.00	1872			
1872				Jan. 8	" "		60.60
Jan. 1	" "		25.00				
" 8	" "		35.65				
			$448.65				$448.65

	Dr.		Boston & Albany Railroad				Cr.
1873				1873			
Sept. 3	To Bal		$.60	Sept. 30	By freight		$2,720.06
" 9	" draft		842.77				
" 22	" "		1,058.14				
" 29	" "		430.21				
Oct. 3	" "		388.34				
			$2,720.06				$2,720.06

[1] Monthly pay of a cutter on a piece work basis.

	Dr.		Boston & Albany Railroad			Cr.
1873				1873		
Oct. 8	To draft		$614.14	Oct. 31	By freight	$1,626.75
" 21	"	"	536.59			
" 28	"	"	260.10			
Nov. 3	"	"	215.92			
			$1,626.75			$1,626.75

	Dr.					
1873				1873		
Nov. 11	To draft		$275.94	Nov. 29	By freight	$844.96
" 19	"	"	226.95			
" 25	"	"	148.57			
Dec. 2	"	"	193.50			
			$844.96			$844.96

	Dr.		McKay Heeling Machine Association			Cr. Boston, Mass.
1874				1874		
Dec. 7	To draft		$242.30	Nov. 30	By bill	$242.30
1875				Dec. 31	" "	243.56
Jan. 4	"	"	243.56	1875		
Feb. 4	"	"	248.28	Jan. 30	" "	248.28
Mar. 10	"	"	274.95	Feb. 27	" "	274.95
Apr. 5	"	"	397.60	Mar. 31	" "	397.60
May 5	"	"	380.25	May 5	" "	380.25

Purchases, 1870–1875

E. & A. Batcheller were buying stock from many firms. They had accounts with: —

> Whittemore Brothers & Co., Boston; for foot power pegging machinery; also other machinery and blacking.
> Calvin W. Hoyt, painter on jobs.
> Jesse Moulton, foundry man at East Brookfield; made presses and cast iron work.
> W. H. Whiting, Express (local).
> E. Howe, Express agent.
> American Sack Co., Fairhaven, Mass.; cotton lining.
> George S. Homer, New Bedford; cotton lining.
> Calvin Foster & Co., Worcester; hardware.
> Mt. Tom Thread Co., East Hampton, Mass.
> Bay State Union Oil Works, Boston.
> Bay State Grease Co., Boston.
> Dunbar, Hobard & Whidden, South Abington, Mass.; makers of nails; slugs for heels.

APPENDICES 255

Brown & Bros., Waterbury, Conn.
A. C. & A. H. Foster Contract Co.; contracting, shipping, including polish, ties and foxing.
P. H. Kellogg, contract account; for finishing bottoms.
J. B. Dewing, foreman; handled thousands of dollars 1869–74; later had contract; bought the last factory from Haskells.
Stoddard & Montague, contractors, after Duncan.
Last Factory 1872–75 (Batchellers owned this, but had separate book-keeping system for it).
Smith & Dane, West Brookfield, Mass.; Batchellers sold lasts to them.
J. Punderford, of New Haven, Conn.; bought lasts.
Hunt, Holbrook & Barber, of Hartford, bought lasts.
Dexter, Stoddard, carpenter account; carpenters for all the buildings.
Charles B. Lincoln.
O. O. Patten.
Pawtucket Braid Co.; linings.
Bay State Needle Co., Worcester; needles.
John D. Lamson.
American Shoe Shank Co., Boston; pressed leather-board shank.
Sumner, Pratt & Co., Worcester; iron shafting.
Leonard, Bundy & Co., Boston.
Bliss & Potter, Brookfield; hardware.
Wood & Light Machine Co., Worcester, machinery.
N. E. Awl & Needle Co., West Medway.
Willimantic Linen Co., Boston, Mass.
North Brookfield Savings Bank.
Excelsior Printing Co., North Brookfield.
First Congregational Society, North Brookfield.
Donner Kerosene Oil Co., Boston.
McKay Heeling Machine Association, Boston.
Cutter Bros. & Co., Boston.
Albert Hobbs.
Consolidated Wax Thread Sewing Machine Co. of Boston.
American Fire Extinguisher Co. of Boston.
Vitrified Emery Wheel Mfg. Co., Ashland, Mass.
Joseph F. Sargent, Boston; machinist, repairing.
Putnam Machine Co., Fitchburg; engine for main power.
A. M. Howe; for trees.
Barbour Bros., N. Y.; thread, linen.
Grafton Awl Co.
A. & W. Haskell; last makers.
Ebenezer Howe, Spencer, Mass.; cases, boxes.
J. R. Cushman & Son, Amherst; leather board.
Blood & Delane; box makers.
Tuttle & Adams, Boston.
Anthony & Cushman, Taunton; nails.
N. D. Ladd & Co., Sturbridge, Mass., die makers for heels and uppers and sole leather.

XXVIII

Samuel Drew, the Shoemaker Metaphysician

Samuel Drew, known in England as the metaphysician,[1] the "Self-Taught Cornishman," wrote and had published during his life as a shoemaker, two books, one on Immortality of the Soul, and another, a reply to Thomas Paine's Age of Reason, which brought him into notoriety and obtained for him a name as an acute thinker and an able controversialist. In later years, he preached, wrote and published many books, and was complimented by two universities, Aberdeen conferring the degree of A.M., and London requesting him to be put in competition for the Chair of Moral Philosophy.

He started out in life at ten and a half years of age as an apprentice to a master who added farming to shoemaking, and made his apprentice do likewise. At 17, disgusted with privations and the slight knowledge and experience in shoemaking he was getting, he ran away before his contract for nine years was fulfilled. A good natured shoemaker in a village where he stopped in his flight, took him on as a journeyman, evidently neglecting to ask for the lad's credientials as to a completed apprenticeship. Here the lad fell into bad company, and was sent by his father to St. Austell, to work under a master who combined the three somewhat kindred businesses of saddler, shoemaker and bookbinder. His shop was also a regular meeting place for the gossipers of the town and for religious discussion. Drew, meanwhile had become active in religious work and views, and learned to debate in this shop with these men while at work with his master on shoes. From this to his later studies, he was led by reading Locke's essay on the Human Understanding, when a copy of it was brought to the shop to be bound. "This book set all my soul to think," he said later.

Drew continued working industriously at his trade, and filling up all his spare moments by reading such books as came to the shop to be bound, or any others he could borrow from friends. Attracted by one science after another, and finding, as most eager minds do, a charm in each, he finally settled to metaphysics, because as he sometimes shrewdly observed, among other recommendations it has this, that it requires fewer books than other branches of study, and may be followed at the least expense. "It appeared to be a thorny path; but I determined nevertheless to enter and begin to

---

[1] Taken from Winks: Lives of Illustrious Shoemakers, pp. 109-121. Cf. J. Sparkes Hall: Book of the Feet. Sketch of Drew, pp. 163-175.

tread it," he remarks; and adds, "To metaphysics I then applied myself, and became what the world and Dr. Clarke call a METAPHYSICIAN."

By the advice and help of friends he resolved, in January, 1787, to commence business on his own account. His savings at this time amounted to only fourteen shillings. He was therefore compelled to borrow capital, or remain a journeyman. It was not difficult, however, to find a man in St. Austell who was willing to trust the now steady and hardworking shoemaker. A miller advanced him £5 on the security of his good character, saying, 'And more if that's not enough, and I'll promise not to demand it till you can conveniently pay me.' Fortunately for him, at this time Dr. Franklin's 'Way to Wealth' came into his hands, and impressed him deeply with its sage maxims and sound principles of business and thrift. On one maxim, though severe, he often at this time acted literally, 'It is better to go supperless to bed than to rise in debt.' The account which he gives of the hard work and rigid economy, and the good fruits they bore, during his first year's experience of business, is highly creditable to him, and will be best told in his own words: 'Eighteen hours out of the twenty-four did I regularly work, and sometimes longer, for my friends gave me plenty of employment, and until the bills became due I had no means of paying wages to a journeyman. I was indefatigable, and at the year's end I had the satisfaction of paying the five pounds which had been so kindly lent to me, and finding myself, with a tolerable stock of leather, clear of the world.' . . .

An incident which happened about this time will show to what dangers his social disposition and fondness for debate exposed him, and how slight an incident saved him from the snare. He had become enamoured of political matters, and discussed them very vigorously with his customers and others who made his work-room a meeting-place where they might hear and debate the latest news. Sometimes these discussions drew him from home into the house of a neighbor, and so absorbed his time that he found himself at the end of the day far behind in his work, and obliged to sit up till midnight in order to finish it. One night, however, he received a severe rebuke from some anonymous counsellor, which effectually put a stop to this bad habit. As he sat at work after most of the neighbors were in bed, he heard footsteps at the door, and presently a boy's shrill voice accosted him through the keyhole with this sage remark: 'Shoemaker, shoemaker, work by night, and run about by day!' 'And did you,' inquired a friend to whom Drew told the story, 'pursue the boy and chastise him for his insolence?' 'No, no,' replied Drew, who had the wisdom to see that there was more fault in himself than the boy, and had also the moral courage and firmness of character to turn the annoyance to profitable account — 'No, no. Had a pistol been fired off at my ear I could not have been more dismayed or confounded. I dropped my work, saying.to myself, ' True, true, but you shall never have that to say of me again!' Right well did he keep to his resolve, and with what results we shall see.

XXIX

An English Bagman

We have the story of Dr. William Carey, a bagman in England in the last quarter of the eighteenth century, who was also a well-trained shoemaker, like Josiah Field, and tramped about to sell his goods. He served his apprenticeship as a lad, worked as a journeyman for twelve years, meanwhile serving as a school teacher and a village pastor in North Hamptonshire. Though proficient as a shoemaker, his studies made him absentminded enough to forget to fit shoes sometimes, and at others to offer two shoes which were not a pair. To dispose of his shoes in order to get money for more stock was a problem after he assumed the problems of independent master, so when times were bad he was obliged to travel from village to village to dispose of his work and get fresh orders. "Once a fortnight he could be seen walking eight or ten miles to Northampton with his wallet full of shoes slung over his shoulder, and returning home with a fresh supply of leather." After that for forty years he worked and was known as an Oriental scholar, a translator of the Bible, a maker of a dictionary and grammars for the study of the East Indian tongues, but to the end of his life he used to claim that he was "only a cobbler." [1]

XXX

An English Story of the Birth and Training [2] of St. Crispin and His Brother

How Crispianus and his brother Crispine, the Two Sons of the King of Logria (thro the Cruelty of the Tyrant Maximinus) were fain in disguised Manner to seek their Lives Safety, and how they were entertained by a Shooe-maker in Feversham.

When the Roman Maximinus sought in cruel sort to bereave this Land of all her noble Youth, or Youth of Noble Blood; the Virtuous Queen of Logria (which now is called Kent) dwelling in the City of Durovenum, alias, Canterbury, or the Court of Kentish-men, having at that Time two young sons, sought all the Means she could to keep

---

[1] Winks: Lives of Illustrious Shoemakers, pp. 130–137. Cf. Sparkes Hall: History of Boots and Shoes, and Biographical Sketches, pp. 200–201.

[2] The Delightful and Princely History of the Gentle-Craft, pp. 57–65.

them out of the Tyrant's Claws: And in this Manner she spoke unto them.

My dear and beloved Sons, the Joy and Comfort of my Age, you see the Dangers of these Times, and the Storms of a Tyrant's Reign: Who having now gathered together the most part of the young Nobility, to make them Slaves in a Foreign Country, seeking for you also, thereby to make a clear Riddance of all our born Princes, to the end he might plant Strangers in their stead: Therefore (my sweet Sons) take the Counsel of your Mother, and seek in time to prevent ensuing Danger, which will come upon us as suddenly as a Storm at Sea, and as cruel as a Tyger in the Forest; therefore suiting your selves in honest Habit, seek some Service to shield you from Mischance, seeing Necessity hath privileged those Places from Tyranny. And so (my Sons) the Heavens may raise you to your deserved Dignity and Honour.

The young Lads seeing their Mother was so earnest to have them gone, fulfilled her Commands; and casting off their attire, put homely Garments on, and with many bitter Tears took leave of their Mother, desiring her to bestow her Blessing on them.

O my Sons (*quoth she*) stand not now upon your Ceremonies, had I leisure to give you one Kiss, it were something, the Lord Bless you! get you gone, away, away, make hast, I say, let not swift Time overslip you, for the Tyrant is hard by: With that she pushed them out at a back Door, and then set her self down to weep.

The two young Princes, which like pretty Lambs were straying they knew not whither, at length by good Fortune came to Feversham, where before the Day peep they heard certain Shooe-makers singing, being as pleasant at their Notes, as they sat at their Business. . . .

The young Princes perceiving such Mirth to remain in so homely a Cottage, judged by their pleasant Notes, that they were not cloyed with many Cares: And therefore wished it might be their good hap to be harboured in a Place of such great Content.

But standing a long time in doubt what to do, like two distressed Strangers, combating betwixt Hope and Fear, at length taking Courage, Crispianus knocked at the Door: What Knave knocks there? (*quoth one of the Journey-men*) and by and by takes his Quarter-Staff, and opens the Door: being as ready to strike as speak saying, What lack you? To whom Crispianus made this Answer: Good Sir, pardon our Boldness, and measure not our Truth by our Rudeness, we are two poor Boys that want a Service stript from our Friends by the Fury of these Wars, and therefore are we enforced succourless to crave Service in any Place. What have you no Friends or Acquaintance in these Ports to go to (said the Shooe-maker) by whose Means you might get Preferment? Alas! Sir Crispianus, Necessity is despised of every one, and Misery is trodden down of many, but seldom or never reliev'd; yet notwithstanding, if Hope did not yield us some Comfort of good Day, we should grow desperate thro' Distress. That were great pity (said the Shooe-maker) be content, for as our Dame tells our Master, a patient Man is better than a strong Man: Stay, and I will call our Dame, and then you shall hear what she will say. With that he went in, and forth came his Dame, who beholding the Youths said: Now, alas! poor Boys, how comes it to pass that you are out of Service? what! would you be Shooe makers and learn the *Gentle-Craft?* Yes, forsooth, (said they) with all our Hearts. Now, by my troth (quoth she) you do look with honest true Faces, I will entreat my Husband for you, for we would gladly have good Boys; and if you would be Just and True, and serve God, no doubt but you may do well enough, come in my Lads, come in. Crispianus and his Brother, with great Reverence gave her Thanks; and by that time they had stayed a little while, down came the Good-man, and his Wife hard by his Heels, saying, Husband, these be the Youths I told you of, no doubt but in time they will be good Men.

Her Husband looking very wishfully upon them, and conceiving a good Opinion of their Favours at length agreed that they should dwell with him, so that they would be bound for Seven Years. The Youths being contented, the Bargin was soon ended, and so set to their Business; whereat they were no sooner settled, but that great Search was made for them in all Places; and accordingly the Officers came to the House where they dwelt, but by reason of their Disguise, they knew them not, having also taken upon them borrowed Names of Crispianus and Crispine.

They both bent their whole Minds to please their Master and Dame, refusing nothing that was put 'em to do; whether it to was Dishes, scour Kettles, or any other thing whereby they thought their Dame's Favour might be gotten, which made her the readier to give them a good Report to their Master, and to do them any other Service, which otherwise they should have missed; following therein the Admonition of an old Journey-man, who would aways say to the Apprentices:

> However Things do frame,
> Please well thy Master, but chiefly thy Dame.

Now by that time these two young Princes had truly served their Master the space of Four or Five Years, he was grown somewhat wealthy, and they very Cunning in their Trade, whereby the House had the Name to breed the best Workmen in the Country, which Report in the end, preferred their Master to be the Emperor's Shooemaker; and by this Means his Servants went to Maximinus's Court every Day; but Crispianus and Crispine fearing they should be known, kept themselves from thence as much as they could; Notwithstanding at the last, perswading that Time had worn them out of Knowledge, they were willing in the end to go thither, as well to hear Tydings of Queen their Mother, as also seek their own Preferment.

XXXI

A French Story of the Birth and Training of St. Crispin and his Brother [1]

Undoubtedly the first shoemakers who obtained anything like a general reputation were the famous brothers Crispin and Crispianus, who are said to have lived in the third century of our era. These saints have been regarded almost ever since that early time as the tutelary or patron saints of shoemakers, who are, to tell the truth, not a little proud of their romantic title, "the sons of Crispin." We must be careful how we speak of these saints, for it seems to be an open question whether the story of their holy self-denying lives and martyr-deaths be true or false. If the main features of the story be true, they have been greatly distorted by fable. We give the story as it is generally reported.

SS. *Crispin and Crispianus* were born in Rome. Having become converts to Christianity, they set out with St. Denis from that city to become preachers of the Gospel, traveled on foot through Italy, and finally settled down at a little town, now called Soissons, in the modern department of Aisne, about fifty or sixty miles to the north-east of Paris. Here they are said to have devoted their time during the day to preaching, and to have maintained themselves by working during most of the night as shoemakers. This they did on the apostolic model of Paul, who, while he carried on his mission as a preacher, maintained himself by his trade as a tent-maker, that he might be "chargeable to no man." Very little more can be told of the life of these saintly shoe-

[1] Quoted from Winks: Lives of Illustrious Shoemakers, pp. 197-198.

makers than this; but this, surely, is a great deal. The story goes that they suffered martyrdom by the order of Rictus Varus, governor or consul in Belgic Gaul, during the persecution under Diocletian and Maximinus, on the 25th of October, 287. The 25th of October is still kept in honor of these saints in some parts of England and Wales, and in other European countries. The shoemakers of the district turn out in large numbers and parade the streets, headed by bands of music, and accompanied by banners on which are emblazoned the emblems of the craft.

It is difficult, as already intimated, to tell how much of pure legend has been imported into the history of the saints of Soissons. One tradition declares them to have been of noble birth, and to have adopted their humble trade entirely for Christian and charitable purposes. Another story relates how they furnished the poor with shoes at a very low price, and that, in order to replenish their stock, and as a mark of divine favor, an angel came to them by night with supplies of leather; while yet another fable, not very creditable to their morals, avows that Saint Crispin stole the leather, so that he might be able to give shoes to the poor. Hence comes the term Crispinades to denote charities done at the expense of other people. To crown all, it is averred on one authority that after suffering a horrible death by the sword, their bodies were thrown into the sea, and were cast ashore at Romney Marsh. Such tales are worthless, except as indicating the wide extent of popularity the shoemakers of Soissons secured by virtue of their piety and benevolence.

XXXII

THE PIOUS CONFRATERNITY OF BROTHER SHOEMAKERS, FOUNDED BY HENRY MICHAEL BUCH [1] (MIDDLE OF 17th CENTURY)

The founder of this society was Henry Michael Buch, who was known throughout Paris, in his day and long after, as *Good Henry*.

Henry Michael Buch came from the Duchy of Luxemburg, where he had been born, and where his parents, who were day-laborers, had brought him up in a very simple manner. As a child, Buch was remarkably gifted and very pious. He was early apprenticed to a shoemaker, and was accustomed to spend his Sundays and holidays in

---

[1] Cf. William Edward Winks: Lives of Illustrious Shoemakers, pp. 201–202.

public worship or private devotion. During his apprenticeship he began the work of reform among the members of his own craft, for his young heart was grieved to see them living in ignorance and vice. Enlisting the help of the more serious among them in his good work, he endeavored to instruct the apprentices of the town in the doctrines of religion, to draw them away from the alehouses and vicious company, and to persuade them to spend their time in a sensible and profitable manner. Taking the patron saints of the trade for a model, he cultivated habits of self-denial and beneficence, went always meanly clad, abandoned luxuries in food and clothing, and frequently gave away his own garments in order to clothe some poor brother shoemaker. While at Luxemburg and Messen, he lived chiefly on bread and water, so that he might be able to feed the hungry and destitute.

Having removed to Paris, his good deeds soon attracted the attention of Gaston John Baptist, Baron of Renti, who was so much impressed by the shoemaker's simplicity of manner, intelligence, and missionary zeal, that he persuaded Buch to establish in that city a confraternity among the members of his own humble craft for the purpose of instructing them in the principles and practices of a holy life. With a view to strengthen his hands for such a task, the freedom of the city was purchased for him, and means were supplied him for starting in business as a master shoemaker, "so that he might take apprentices and journeymen who were willing to follow the rules that were prescribed them."[1]

Seven men and youths having joined him on these terms, the foundation of his Confraternity was laid in 1645, Good Henry being appointed the first superior.[2]

Two years after this, the *tailors* of the city, who had noticed the conduct of the shoemakers, and had been delighted with the goodly spectacle presented in their happy and useful lives, resolved to follow the example. They borrowed a copy of the rules, and started a similar society in 1647.

These brotherhoods, but notably those of the shoemakers, were spread through France and Italy, and were the means of doing an immense amount of good among the members of the two crafts.

The rules of the fraternity founded by Buch were assimilated to certain monastic orders. They enjoined rising at five o'clock and meet-

[1] Butler's "Lives of the Primitive Fathers, Martyrs, and Saints," 1799, p. 532.
[2] This society flourished until the outbreak of the French Revolution, 1789, when it was suppressed.

ing for united prayer before engaging in work, prayers offered by the superior as often as the clock strikes, at certain hours the singing of hymns while at work, at other times silence and meditation; meditation before dinner, the reading of some devotional work by one of the number during meals; a *retreat* for a few days in every year; assisting on Sundays and holy days at sermons and "the divine office"; the visitation of the poor and sick, of hospitals and prisons; self-examination, followed by prayer together at night and retiring to rest at nine o'clock.

XXXIII

The Shooe-maker's Glory

BEING

A Merry Song in the Praise of Shooe-makers, to be sung by them every Year on the 25th of October

To the Tune of, *The Tyrant*, &c.

In the Praise of the
  Shooe-makers Trade we'll write,
A Merry Song is to be sung
  on October's Twenty fifth Night.
For without the Shooe-maker
  we shall go cold of our Feet,
To preserve the Gentle-Craft
  therefore it is meet.
Then sing Boys, and drink, Boys,
  and cast Care away,
For the honour of Shooe-makers,
  we'll keep Holiday.

To add the more Lustre unto due Merriment
  Our Ancestors came of a Royal Descent:
Crispiana, Crispinus, and Noble St. Hugh,
  Were all Sons of Kings, this is known to be true:
  Then sing Boys, and drink Boys, &c.

Moreover I wou'd have you thus much understand,
  That the chiefest gay Ladies, and Lords of our Land,
To the bonny Shooemakers beholding must be:
Take them from the highest to the lowest Degree:
  Then sing Boys, and drink Boys, &c.

And now for St. Hugh, and fair Winifred's sake,
A jovial Bout of it, we purpose to make;
In the Gulf of Oblivion let Sorrows be drown'd,
Whilst we in good Fellowship merrily drink round:
    Then drink Boys, and sing Boys, &c.

Here's a Health to the Muses, which furthers Delight,
And helps us to pass away long Winter Nights:
With Songs and with Pastimes, as the Season doth require,
Whilst we soak our Noses and sit by the Fire;
    Then sing Boys, and drink Boys, &c.

The next cordial Health, to speak as I think,
Shall be to the Brewer that makes us good Drink;
And to the good Butchers, that kills us good Meat,
That's toothsome and wholesome for Christians to eat,
    Then sing Boys, and drink Boys, &c.

Here's to the bonny Weavers, and Glovers also,
For they are our Neighbours and Men that we know;
And to Vulcan the Blacksmith that blowest the Bellows,
For he is accounted the King of Goodfellows;
    Then sing Boys, and drink Boys, &c.

Here's to the Taylor, that never meant harm,
For he makes us Cloathing to keep our Bones warm,
And a Health to the Tanner that dresseth our Leather,
For they are the Men that must hold us together.
    Then sing Boys, and drink Boys, &c.

And now to conclude all, and finish my Song,
  Let's drink up our Drink, and do no Body wrong:
  'Tis late in the Night, Sirs, therefore let us pay
  Our Reckoning, and then we'll be jogging away;
  Another Time when we do meet here again,
  We'll make a merry Bout for an Hour or twain.

    Reprinted in The Delightful and Princely History of the Gentle-Craft, pp. 165–166.

XXXIV

Brothers of St. Crispin

    Brothers of St. Crispin were so well known in Elizabethan times that Shakespeare ventured to devote several lines to them at a most vital point in King Henry Fifth's address to the leaders of the English forces just before the Battle of Agincourt (Act V, Sc. 3), knowing he had many shoemakers in his audience.

This day is called the feast of Crispin.
. . . .
And Crispin Crispian shall ne'er go by,
From this day to the ending of the world,
But we in it shall be remembered.
We few, we happy few, we band of brothers.

Unfortunately for the shoemaker fame, the thousands of boys and girls in High School today and the great majority of grown men and women who hear the lines on the stage each year, have no connotation with St. Crispin, so brief was the revival of the Order of St. Crispin and so entirely local were its interests and effects.

XXXV

New Help[1]

No member of this order shall teach or aid in teaching any part or parts of boot or shoe making, unless this Lodge shall give permission by a three-fourths vote of those present, and voting when such permission is first asked. Provided this article shall not be so construed as to prevent a father teaching his own son. Provided, also, that this article shall not be so construed as to hinder any member of this organization from learning any or all parts of the trade.

XXXVI

Ritual of the Degree of . . . in the Order of the Knights of St. Crispin, Temple of . . . Boston, 1870

Obligation

You each and all solemnly and sincerely pledge yourself, your sacred word and honor, that you will, under no circumstances, divulge any of the secrets of this Temple to any person whom you do not *know* to be a member of the Temple of which you are a member, and whose standing is good (except your religious confessor); that you will not make known any signs, pass-words, tests, or any other work of the Temple. You also pledge yourself to bear true and faithful allegiance to the Order of the Knights of St. Crispin, and obey and enforce its con-

---

[1] Constitution of the Subordinate Lodges of the Order K.O.S.C., Article X.

stitutional rules and obligations, to the best of your power and ability, by all proper means; and you also agree to do all you can to persuade all true and loyal Crispins to join this Temple, and unite in this work of protecting and elevating labor. All this you promise freely of your own accord, and without any mental reservation whatsoever. You also agree to be governed by the will of the Temple expressed in the manner provided in the Constitution. This obligation you agree to keep inviolate, as long as these Temples exist.

All this you promise, on honor before God and these witnesses, who will bear swift witness against you should you prove false.

XXXVII

The Crispins at Burrell & Maguire's Factory

Workmen in Randolph say that the Crispin demands threatened the profits of Burrell & Maguire. Business men say that when the Australian trade [1] died out in the late 60's, that the Burrell & Maguire market "went flat." A third explanation which is more significant, is found in the fact that Mr. Maguire was undeniably a Copperhead and his son James Frank Maguire, the former American Consul at Melbourne, was one also. Both men at the close of the war found themselves in the midst of strong Union sentiment which may have vitally affected their business. The following item from a bit of faded yellow newspaper has come into my hands:

The Pirate Shenandoah. A private letter from Melbourne, Feb. 24, thus speaks of the visit of the rebel pirate Shenandoah to that place:

The Confederate war pirate Shenandoah has been in Hobson's Bay about a month. The officers have been lionized and fêted all over the town and country. You would scarcely believe there were so many "secesh" here as there are. They gave them a public dinner here and in Bellarat. J. F. Maguire, late American Consul, showed them the most attention. We tried to blow her up, but could not get a chance.

There is to me a fourth explanation, *i.e.*, that both members of the firm, and their agent in Australia had all become wealthy and interested in other investments by 1872, so that the pressure of the Crispins at just the opening of hard times in 1873, was the straw that decided the firm to retire from the shoe business since they could do it with comfortable fortunes.

[1] I presume they mean the early monopoly of it.

SOURCES

A. Primary Sources.[1] Oral

Barnes, Dow	Frenchtown, Pa.
Batcheller, Francis	North Brookfield, Mass.
Belcher, Fred L.	Randolph, Mass.
Belcher, Joseph	Randolph, Mass.
Belcher, Mary E.	Randolph, Mass.
Belcher, Stephen	Randolph, Mass.
Brown, Richmond	South Hanson, Mass.
Brown, Lucy	South Hanson, Mass.
Cole, Abbie	Warren, Rhode Island
Cooper, James H.	Brockton, Mass.
Cox, Samuel	Lynn, Mass.
Crocker, Henry P.	Raynham, Mass.
Felch, Henry	Natick, Mass.
Felch, Isaac	Natick, Mass.
Field, John H.	Randolph, Mass.
Fletcher, Jerome C.	Littleton, New Hampshire, and later of Brockton, Mass.
Gilmore, Cassander	Raynham, Mass.
Holmes, Sumner	North Brookfield, Mass.
Jones, Asa	Nantucket, Mass.
Leach, George Myron	Raynham, Mass.
Littlefield, Cyrus	Avon (then East Stoughton), Mass.
Malloy, James	Randolph, Mass.
Martin, Benjamin	Bristol, Rhode Island
Puffer, Loren W.	Brockton, Mass.
Spinney, Benjamin F.	Lynn, Mass.
Tirrell, Henry A.	East Weymouth, Mass.
Tucker, Nathan	Avon, Mass.
Twitchell, Henry W.	Brookfield, Mass.
Wales, Charles W.	Randolph, Mass.
Whitcomb, Joseph	Holbrook (then East Randolph), Mass.
White, Samuel	Randolph, Mass.
Worthley, Hiram S.	Strong, Me., and later of Concord, Mass.

B. Primary Sources. Manuscript

1. Allen Papers in American Antiquarian Society Library, at Worcester, Mass. Letters of Captain Thomas Willson.
2. Batcheller Papers. Account books of Ezra and Tyler Batcheller and successors, written from 1832 to 1875. At present, these are stored in North Brookfield, Mass., in the guardianship of Mr. Francis Batcheller.

---

[1] All of these people quoted as primary sources were born from eighty to one hundred years ago. Each of them has added definitely to the information given in these pages.

SOURCES

3. Belcher Books. Exercise Book of 1793 and an account book from 1807–1844. Written in Brookville (South Randolph), Mass., by Ebenezer Belcher, and given by his son's widow to the Harvard University Archives.
4. Breed Books. Account books of Amos Breed of Lynn, Mass. Written 1763–1796, and now preserved in the Lynn Historical Collection.
5. Howard and French Books. Account books, including day books and ledgers written from 1849–1855 in the offices of the Howard and French Company of Randolph, Mass. Now owned by Blanche E. Hazard.
6. Kimball and Robinson Papers. Written in North Brookfield, Mass., 1858. Owned in 1908 by Henry Emmons Twitchell and stored in his attic in North Brookfield, Mass.
7. Reed Account Book. Written in 1729–1763 by George Reed of Dighton, Mass. Now owned by Miss Julia Gilmore of Raynham, Mass.
8. Reed Papers. Account books, letters, bills, receipts, bills of lading and invoices belonging to Harvey and Quincy Reed of South Weymouth, Mass., covering the years 1809–1838. Preserved by Quincy Reed, Jr., of South Weymouth, Mass., in 1908.
9. Robinson and Company Account Book of 1848–1854. Written at Lynn, Mass., and preserved in the Lynn Historical Society Collection.
10. Skinner and Ward Store Books. Written from 1813–1815 in North Brookfield, Mass.
11. Southworth Papers. Accounts kept by Jedediah Southworth in Stoughton, Mass., 1811–1814. Owned now by Loren W. Puffer of Brockton, Mass.
12. Tobin Letter. Written in 1916 about the History of the order of St. Crispin. Owned by Blanche E. Hazard.
13. Twitchell Account and Order Books for 1861–1865, kept by Henry E. Twitchell in North Brookfield, Mass.
14. Wendell Papers. Letters, bills and receipts, account books and memoranda recording the business transactions of John Wendell in Portsmouth, New Hampshire, including the years 1802–1804. Owned and carefully filed by Mr. Barrett Wendell of Boston, Mass.
15. White and Whitcomb Account Book. Kept in East Randolph, Mass., in 1833. Preserved by Joseph Whitcomb in 1908–1912 at his Raynham home.
16. Wilson Account Book for 1846. Written and owned by Mr. Henry Wilson of Natick, Mass. Now preserved by Mr. Louis Coolidge, Treasurer of the United Shoe Machinery Company, Boston, Mass.

C. PRIMARY SOURCES. PRINTED

1. *Autobiographical, biographical and historical.*
    Bryant, Seth: Shoe and Leather Trade for one hundred Years. Ashmont, Mass., in 1891.
    Howe-Singer-Leavitt Machine Controversy. First hand reports, printed in the Sewing Machine Journal of April 25 and May 25, 1904; February 10, July 10, and July 25, 1911.

Johnson, David N.: Sketches of Lynn or The Changes of Fifty Years. Lynn, Mass., T. P. Nichols, 1880.
Johnson, Edward: Wonder Working Providence of Sion's Savior in New England. 1654.
Larcom, Lucy: New England Girlhood. Boston, Houghton Mifflin Company, 1884.
Larcom, Lucy: Hannah Binding Shoes (a poem). Boston, Houghton Mifflin Company, 1884.
Porter, J. W.: Interviews with Quincy Reed in 1885. Published in the Bangor Historical Magazine, Vol. I, August, 1885.

2. *Documentary.*
Charter of the Boston Shoemakers of 1648. Reprinted in the Records of the Colony of Massachusetts Bay in New England, Vol. III, p. 132.
Constitution of the Knights of St. Crispin. Printed in the University of Wisconsin Economic Series, Bulletin No. 355, The Knights of St. Crispin, written by Don D. Lescohier. Madison, Wisconsin, July, 1910.
Speeches of the Chief Knight and of the International Grand Scribe of the Order of St. Crispin, 1870 and 1872. Reprinted in the above bulletin.

3. *Public Records.*
Annals of Congress. Tariff Debates, 1834.
Massachusetts Census of 1855, 1865, 1875. (Tables.)
Suffolk County Records for 1757.
United States Census of 1860. (Tables.)

4. *Newspapers and Advertising Cards.*
(a) Cards to advertise sailing of vessels. (Collection preserved by Augusta B. Wales of Randolph, Mass.)
(b) Files[1] of the Boston Gazette, 1754–1764; Charleston Gazette, 1783; City Gazette (of Charleston, S. C.), 1790; Hampshire Gazette, 1789; Hide and Leather Interest, May, 1869; Lynn Reporter, 1857; Massachusetts Spy, 1771, 1773; Middlesex Gazette, 1795; Newburyport Herald, 1857; Pennsylvania Gazette, 1771; Randolph Transcript, 1857; Salem Gazette, 1795; Salem Mercury, 1787; South Carolina Gazette, 1784; Worcester Gazette, 1798, 1799.

D. SECONDARY SOURCES. PRINTED

1. Bacon, Oliver N.: History of Natick from its Settlement in 1651 to the Present Time. Boston, Damrell and Moore, 1856.
2. Barber, John Warner: Historical Collections of every town in Massachusetts. Worcester, Dorr, Howland and Company, 1839.
3. Bücher, Karl: Die Entstehung der Volkswirtschaft. Third edition, Leipzig, 1900. English translation by Wickett. Boston, Henry Holt and Company, 1912.

[1] This list includes only the papers quoted, as typical of the files consulted.

SOURCES

4. Commons, John K.: American Shoemakers. (Quarterly Journal of Economics, November, 1909.)
5. Delightful, Princely and Entertaining History of the Gentlecraft of Shoemakers. London, 1725.
6. Hall, J. Sparkes: History of Boots and Shoes, and Biographical Sketches. New York, William H. Graham, 1847.
7. Hurd, Duane H.: History of Middlesex County. 3 vols. Philadelphia, Pa., J. M. Lewis and Company, 1890.
8. Kingman, Bradford: History of North Bridgewater. Boston, 1866.
9. Lescohier, Don D.: The Knights of St. Crispin. Published in University of Wisconsin Series, Bulletin No. 355, Madison, Wisconsin, 1910.
10. Lewis and Newhall: History of Lynn. Boston, Samuel W. Dickinson, 1844.
11. Proceedings of the 150th Anniversary of the First Congregational Church, Randolph, Mass. Published in Randolph, 1893.
12. Rehe, Carl: Beitrag zur Geschichte der mechanischen Schuhfabrikation, pp. 185–214 (in Beitr. z. Gesch. der Technik u. Industrie), vol. 3, Berlin, 1911.
13. Roberts, James A.: New York in the Revolution as Colony and State. Records discovered and arranged by James A. Roberts. Albany, N. Y., 1897.
14. Temple, Josiah H.: History of North Brookfield, Mass., preceded by an account of Old Quabaug. Brookfield Records 1686–1783. Published by the town, 1887.
15. United States Department of Labor: Wages and Hours of Labor in the Boot and Shoe Industry. Bulletin No. 134 of the U. S. Bureau of Labor Statistics.
16. Unwin, George: Industrial Organization in the Sixteenth and Seventeenth Centuries. Oxford, Clarendon Press, 1904.
17. Winks, William E.: Lives of Illustrious Shoemakers. New York, Funk and Wagnalls, 1883.

INDEX

INDEX[1]

Roman figures refer to the preface; Arabic to the text.
Whenever a subject is treated in a footnote as well as in the text, the text alone is indicated.
Subjects not found among the headings should be sought among the subheads under Human element and Stages of production.
The following abbreviations are used:
n. 1, n. 2, etc., indicates the number of the note.
Ho means Home stage.
Ha means Handicraft stage.
D, D1, etc., means Domestic stage, the figure indicating the Phase.
F, F1, etc., means Factory stage, the figure indicating the Phase.
KSC means Knights of St. Crispin.

ABINGTON, a shoe town,
  in early days, 178.
  in D, 52, 72, 100 n. 1, 209.
  in F, 117–119, 141, 147 n. 3, 153.
Alden family of Randolph, entrepreneurs throughout D, 18 n. 1, 19, 45, 46, 50, 99 n. 1, 108 n. 1, 117 n. 1.
Apprentices, see Human element, apprentices.
Australia, sale of Massachusetts made shoes in,
  in D3, 81, 83.
  in F1, 98–101, 103, 107, 110 n. 1, 267.
Avon (formerly E. Stoughton), a shoe town, 48 n. 2, 50–52, 62.

BACON, Oliver, *History of Natick*, 220.
Baker, Virginia, *The Answer to Abiel Kingsbury's Prayer*, 6 n. 1.
Baltimore, 99 n. 1, 108 n. 1, 196.
Barber, John W., *Historical Collections of Massachusetts*, 66, 207–212.
Batcheller family of North Brookfield, entrepreneurs throughout D and in F, 22, 49 n. 2, 50, 55–59, 63, 113, 116 n. 1, 122, 123, 200–207, 246–255.

Belcher, Ebenezer, a shoemaker-farmer of D2, 20 n. 2, 27, 45–48, 52.
Blake, Lyman R., inventor of the McKay machine, v n. 2, 111, 117, 245, 246.
Boston,
  in Ho and Ha, 10–14, 174, 177, 182, 189.
  in D, 28, 29 n. 3, 67, 135, 196, 209, 211.
  in F, 99, 103, 105–109, 119, 141, 150, 151, 154.
Braintree, parent town of Randolph, 46, 176–178, 182, 208.
  See also Randolph; S. Randolph.
Breed family of Lynn, entrepreneurs of D1, 28 n. 1, 37–39, 79, 179 n. 2, 185–187.
Bridges, Edmund, English shoemaker who came to Lynn in *1635*, 23, 131.
Bridgewater (including Joppa), 52, 59, 61, 177, 182, 209.
  See also Brockton (North Bridgewater); S. Bridgewater.
Brockton (formerly North Bridgewater), a shoe town,
  early conditions in, vi n. 1, 17, 19, 20, 179 n. 1, 180, 181.

---

[1] This index is the work of Mr. Karl W. Bigelow, Instructor in Economics at Cornell University.

INDEX

in D, 20, 67 n. 2, 73, 77, 91, 135 n. 2, 138, 140, 209.
in F, 99, 118, 119, 139, 141, 144 n. 1.
See also Bridgewater; N. Bridgewater.
Brogans specialized in by certain localities, 80, 81, 114, 139.
Brookfield (formerly Quabaug), a shoe town,
early conditions in, 21, 22, 179 n. 1, 181, 182.
in D, 22, 49 n. 2, 50, 55–59, 72, 81, 133, 134, 174–176, 200, 210, 212.
in F, 98, 113–117, 122, 123, 246.
Brown, Lucy (Mrs. Richmond), a contented rapid stitcher of F, 140, 141.
Bryant, Seth, wholesale shoe dealer of D2, 59–63, 117–119.
Shoe and Leather Trade for One Hundred Years, a source of information regarding Ha and early D, 14 n. 2.
See also Mitchell and Bryant.
Buch, Henry Michael, founder of the Pious Confraternity of Brother Shoemakers, 17th century, 262–264.
Bücher, Karl, confirmation of views of, vii, 176 n. 2.
Burrell and Maguire's factory at Randolph, 99, 100 n. 2, 101, 110 n. 1, 153, 267.

CALIFORNIA, 72, 81, 83, 95, 98–101, 103, 107.
Campello, see Brockton.
Carey, Dr. William, an English bagman, 258.
Carver, T. N., *Organization of Rural Interests* (in Yearbook of the Dept. of Agriculture, *1914*), 177 n. 5.
Central shop, see Stages of Production (D2, D3, F1) shops in, central.
Charleston (S. C.), a distribution point for Massachusetts made shoes, 35.
Chicago, 104 n. 4, 151, 152.

Clark, Amasa, agent for Howard and French for local trade in D3, 45 n. 4, 83, 84, 221–225.
Cochato, see Randolph.
Commercial life, effect of on economic life, 16.
Commons, John R., writings of, referred to, 142 n. 1, 143 n. 1.
Connor, Nichols, inventor of binder, 79.
Cox, Samuel, of Lynn, marginal worker of D, 144 n. 2.
Crispianus, brother of St. Crispin, 130, 258–262, 264.
Crispins, see Human element, KSC.
Crocker, Henry P., source of information regarding transition to F, 92 n. 1, 115.
Cummings, S. P., International Grand Scribe of KSC, 154–156.
Customers, see Stages of Production (Ha, D3), customers.

DAGYR, John Adam, Welsh shoemaker who came to Lynn in *1750*, 29, 78, 131.
Daniels, Newell, organizer of KSC, 145, 147, 152, 153.
Danvers, 72, 210, 212.
Delightful, Princely, and Entertaining History of the Gentlecraft, The, 169, 170, 258 n. 2, 265.
Deming and Edwards, North Brookfield firm, 58, 59, 82 n. 2.
Dighton, 174.
Drew, Samuel, shoemaker metaphysician of England, 133, 134, 256, 257.

EAST MIDDLEBORO, see Middleboro.
East Stoughton, see Avon (present name of town).
Economic conditions of early Massachusetts, brief summary of, 15–23.
Economic life, affected by commercial and industrial conditions, 16.
Eddy and Leach, typical partnership of capitalist and shoemaker in D3, 71, 72, 115, 139.

England,
  industrial history of, 26, 27.
  leather imported from in D, 30.
Essex county, a centre for boot and shoe industry, 67, 81, 113, 210-212.
Europe, evolution of industry in, 15.

FACTORIES, see Stages of Production (D3, F1, F2), factories in, Domestic and Factory stages.
Felch family of Natick, early entrepreneurs of D2, 53, 54, 217 n. 2.
Field, Josiah, the Randolph bagman, 11, 135, 258.
Financial history of four Massachusetts shoe centres, early, 176-183.
Fletcher, Jerome C., source of information regarding Ho, 5 n. 1, 150 n. 1.
French, see Howard and French.

GEORGETOWN, 72.
Gilds, see Human element, gilds.
Gilmore, C. and A. H., of Raynham, factory owners, 92 n. 1, 102, 110-112, 114, 230, 231.
Grafton, 55, 72, 200, 201, 209.

HALL, J. Sparkes, *History of Boots and Shoes*, 128, 256 n. 1, 258 n. 1.
Hathaway, Paul, belated itinerant shoemaker, 134, 135, 138.
Haverhill, a shoe town,
  in D, 78, 83, 190, 198, 210, 212.
  in F, 105, 112.
Hemenway, Augustus, developer of South American trade, 62, 63.
Holbrook, 49.
Holliston, 59, 208, 211.
Hopkinton, 72, 208, 211.
Howard, Daniel S., creator of the "good low-priced shoe," 119.
Howard and French's factory at Randolph, 83-92, 99 n. 1, 101, 119, 217 n. 1, 221-230.
Howe, Elias, inventor of the sewing-machine, v n. 2, 93.

HUMAN ELEMENT in boot and shoe industry, 127-156.
apprentices,
  in old countries, 128, 256, 258, 260, 263.
  in Ha, 10-12, 132.
  in D1 and D2, 25, 32, 33, 47, 134, 135, 193, 201.
  scarcity of in D3 and F, 86, 95, 143, 154.
bagmen, 128, 135, 258.
Boot and Shoe Workers' Union, The, developed in F2, 156.
Boston Gild of *1650*, The, a shoemakers' organization, 142, 143, 170-173.
Brothers of St. Crispin, The, an English gild of Shakespeare's day, 265, 266.
child labor, 70, 80.
  See also Human element, young people as shoeworkers.
"closed shop,"
  not attempted by Boston Gild of *1650*, 142.
  rise of feeling for in F1, 144-146, 148, 149, 154, 155.
cobblers, see Human element, shoe workers.
Company of Shoemakers, The, see Human element, Boston Gild.
cordwainers, see Human element, shoe workers.
division of labor, see Stages of production (Ha, D1, D2, D3, F1, and F2), specialization.
family as working unit, 70, 71.
Federal Society of Journeymen Cordwainers, The, an organization never existing in Massachusetts, 142 n. 1.
foreigners instruct American shoemakers in craft, 28, 29.
gilds, very few in Ha, 10.
grievances in F1, 140, 141.
  See also Human element, strikes.
hours of employment, 88, 98, 112.
immigrants as apprentices, 32.

HUMAN ELEMENT (*continued*)
itinerant cobblers, *see* Human element, shoe workers, itinerant.
journeymen,
 in old countries, 133, 135 n. 2, 256–258, 263.
 in Ho and Ha, 6, 10, 12, 128, 142, 185, 186 n. 3.
 in D, 25, 33, 34, 47, 80, 134, 193.
 in F, 141, 143–145.

Knights of St. Crispin, The, a shoe workers' organization existing *1868–1874*, 142–156.
 accomplishments of, 156.
 attitude of, towards members who became foremen and entrepreneurs, 147 n. 1, 148, 152, 153.
 Chief Knight's speech quoted, 151, 152.
 Constitution of, referred to and quoted, 145, 146, 148 n. 2, n. 3, 149 n. 1, n. 2, n. 3, 266.
 decline of, 155, 156.
 demands of, 145.
 discussed by old shoe workers, 140.
 discussion of politics and religion taboo in lodges, 149.
 dues of, troubles in connection with, 147–149.
  See also Human element, KSC, finances.
 duties of members, 148, 149.
 effect of name of, on shoe workers, 145.
 effects of, on boot and shoe industry, 153–155, 267.
 enthusiasm, lack of a fatal defect, 147, 150–153.
 finances of, a great problem, 150–152, 155.
  See also Human element, KSC, dues.
 inflexibility of views, 153.
 instability in action, 153.
 internal dissension, 151.
 membership, 145–147.
  new members, endeavors to win, 149.
  not taken very seriously, 153, 154.
  number of lodges in Massachusetts, 147.
  numbers of Massachusetts shoe workers joining, 145.
 objects of, 145–147.
 panic of 1873's effect on, 156.
 problems of industry not appreciated by, 141, 142.
 problems of Massachusetts lodges, 150–153.
 Proceedings of quoted, 145.
 requirements for membership in, 148.
 rites of, 147, 148, 266, 267.
 rules of, inadequate, 147.
 scabbing, attitude towards, 146, 148.
 sources of information concerning, 145, 146.
 strike expenses of, failure to meet, 149.
 strikes under auspices of, *see* Human element, strikes.
 success only temporary, 153.
 unskilled workers opposed by, 143, 146, 148, 149, 154, 155, 266.
  See also Human element, shoe workers, unskilled.
 vows of, 147, 148, 266, 267.
 wages, attitude of, towards, 146, 155.
 weakness of Executive Committee of, 152.

labor, *see* Human element, shoe workers.
labor problems,
 rise of, 125.
 before organization of KSC, 141, 142.
labor-saving, 95.
master shoemakers,
 in old countries, 133, 260, 263.
 in early Massachusetts, 128, 142.
negroes as apprentices, 32.
order of St. Crispin, The, a European gild of the 16th century, 145.

INDEX 279

HUMAN ELEMENT (*continued*)
Pious Confraternity of Brother Shoemakers, The, a brotherhood of the mid-17th century, 262–264.
regularity of work, an object of KSC, 146.

shoe workers,
Civil War brings out old, 117.
Civil War's effect on status of, 98, 124.
characterized, 127–131.
Chinese imported for use as, 143, 150, 154, 186.
comparison of those in Ha with mediæval shoemakers, 10.
competition among, growth of, in Ha, 11.
competition for,
in D2, 44.
in D3, 144.
reduced with advent of machinery in F, 144.
contact of, with people of all classes, 130.
contract labor of, in Batcheller factory, 255.
domestic,
in D, 52, 53, 100, 198, 227–230.
in F, 109, 116, 124 n. 1.
early English forerunners of those of New England, 128.
economy in employment of, necessity for, in D3, 90, 95.
exemption of, from military service during Revolutionary War, 192, 193.
Factory Stage, effect of, on old, 116.
failure of, to appreciate entrepreneurs' problems in F1, 141, 142.
farmers as, in Ho, 6, 84.
grouping of, best in larger towns in Ha, 12.
ideals of, 130.
importance of, in D3, 75.
influence of old customs and legends on, 128, 130.

itinerant, in Ho, 5, 25, 134, 135, 142.
as lawyers, 132.
little known of, 127.
machinery, effect of, on, 98, 143–145.
numbers employed, in *1837*, 54 n. 2, 207–210, 212; in *1846*, 69; in *1854*, 220; in *1855*, 104; in *1856*, 108; in *1860*, 112, 113; increased greatly in F2, 125.
opportunities of, for thought and study, 129, 130.
as politicians, 128, 129, 132, 133, 257.
as preachers, 128, 129, 256, 258.
as readers, 129, 130, 132–134, 256.
rise of many to rôle of entrepreneur, 257.
See also Stages of production (D1, D2, D3, and F1), entrepreneurs.
scarcity of, during Civil War gave impetus to use of machines, 124.
as scientists, 129.
skill of high order required in factories, 131 n. 1.
skilled, less in demand in D2, 52.
study of individuals as types, 131–141.
success of, in other fields, marked, 128, 129.
superior intelligence of, in general, 128, 129.
supply of, in D1, 32, 33.
as teachers, 138, 139, 173, 258.
as thieves occasionally, 189, 190.
troubles of, during transition from Ho to Ha, 8, 9.
unemployment of, in F1, 106.
unorganized in *1875*, 156.
unskilled, 42, 52, 74, 143–145.
vacations taken, for farming, 84; for summer recreation, 108.
women as, *see* Human element, women.
as writers, 129, 130, 133, 256.
See also other sub-headings under Human element.

HUMAN ELEMENT (continued)
  Society of Master Cordwainers, The, an organization never existing in Massachusetts, 142 n. 1.
  strikes, 146 n. 1, 147, 149–152.
  United Beneficial Society of Journeymen Cordwainers, an organization never existing in Massachusetts, 142 n. 1.
  wages,
    in Ho and Ha, 142, 173, 186, 187.
    in D1, 37, 198.
    in D2, 43, 46, 144.
      methods of payment of, 46, 48, 52.
    in D3, 69, 84–89, 217, 218.
      cash payments after *1845*, 85.
      daily, 87, 88, 115.
      no weekly pay-roll, 85.
      for piece-work, 86–89, 115, 227–230.
      for women, 86, 229.
      survivals of truck system of payment, 84, 85.
    in F1, 140, 231–234, 237.
      advances in, meant loss to entrepreneurs, 141.
      attitude of KSC towards, 146, 155.
      effect of introduction of machinery on, 144, 145.
      inferior workmen could reduce, 144.
      of old retired shoemakers during Civil War, 117.
      "subsistence wages" fought by KSC, 146.
      trial of contract system in connection with, 122.
  women as shoe workers,
    in D1, 25.
    in D2, 47 n. 2, 52, 54 n. 2, 207, 212.
    in D3, 69–71, 114, 140, 220, 227, 229.
      irregular employment of, 100.
      wages of, 86, 229.
    in F1, 98, 101, 104, 113, 119, 124 n. 1, 140.

  effects of Civil War, 98, 117.
  in a modern factory, 160, 162, 163.
  work, amount of, available in D3, 87, 229, 230.
  young people as shoe workers, 100 n. 2, 121 n. 1, 135, 139, 143.
  See also Human element, child labor.

Hyannis, 100 n. 1, 137.

INDUSTRIAL life, effect of, on economic life, 16.
Iron industry, connection with boot and shoe industry at Brockton, vi n. 1.

JOHNSON, David N., *Sketches of Lynn*, v n. 2, 73, 75, 120, 121 n. 1.
Johnson, Edward, *Wonder Working Providence of Sion's Savior*, 9, 170, 171, 183.
Jones, Asa, pioneer domestic worker of Nantucket, 100 n. 1, 137, 138.
Joppa, *see* Bridgewater. (Joppa was a part of East Bridgewater which is now called Elmwood.)
Journeymen, *see* Human element.

KERTLAND, Philip, English shoemaker who came to Lynn in *1635*, 23, 131.
Kimball and Robinson's factory at Brookfield, 114–116, 231–235, 266 n. 1.
Kingman, Bradford, *History of North Bridgewater*, 18 n. 2, 180 n. 1.
Knights of St. Crispin, The, *see* Human element, KSC.

LABOR, *see* Human element, shoe workers.
Larcom, Lucy, poem of, about shoebinders, v n. 2, 131 n. 1.
Lawrence, 190 n. 1.

INDEX 281

Leach family of S. Bridgewater, E. Middleboro, Raynham, and Brockton, shoemakers of Ha, D and F, 115, 138, 139.
See also Eddy and Leach.
Leather, see Stages of Production (Ho, Ha, D1, D2, D3, F1, F2), leather.
Lescohier, Don D., *The Knights of St. Crispin, 1867–1874*, 145 n. 3, 147 n. 2, 148 n. 1, 150.
Lewis and Newhall, *History of Lynn*, 183 n. 1.
Lincoln, Ephraim, one of the first entrepreneurs of D2, not a shoemaker, 48, 49.
Littlefield family of Avon, typical entrepreneurs of D2, 50–53, 62.
Lynn (formerly Saugus), a preëminent shoe town,
 early conditions in, 9, 11, 12, 22, 23, 131, 142 n. 1, 182, 183.
 in D1, 22, 28–31, 34, 39, 40, 188, 195, 197.
 in D2, 50, 59, 63, 210, 212.
 in D3, 72, 73, 75, 77–81, 83, 84, 91, 144 n. 2.
 in F, 98, 101, 105, 106, 109, 112, 114, 119–121, 144 n. 1, 147, 150.

MACHINERY, Human element, shoe workers, machinery; Stages of Production (D3, F1, F2), machines.
Maguire, see Burrell and Maguire.
Maine in F2, 144 n. 1.
Markets, see Stages of Production (Ho, Ha, D, D1, D3), market.
Massachusetts Bay, Records of Colony of, 171–173.
McKay, Col. Gordon, capitalist who purchased from Lyman R. Blake the McKay machine, v n. 2, 111, 121, 245, 246.
See also Stages of Production (F1 and F2), McKay machine.
Middleboro, a shoe town of D, 52, 71, 72, 134, 135, 138, 139, 209.
Middlesex, 34.

Middlesex county, a centre for boot and shoe industry, 81, 208, 211.
Milford, 72, 209.
Milwaukee, origin of KSC in, 145.
Mitchell and Bryant, first wholesale boot and shoe house in Boston, 59–63, 203.

NANTUCKET, shoe-manufacturing in, in Ha and D, 100 n. 1, 137, 138, 179 n. 1.
Natick, a shoe town,
 in Ha, 54.
 in D, 53, 54, 68–72, 80, 208, 220.
New Bedford, 81.
New England,
 boot and shoe manufacture in, in *1860*, 113.
 contemporary account of trades in, in *1650*, 170, 171.
New London (Conn.), 189.
New York City,
 in D, 72 n. 2, 136, 192.
 in F, 104 n. 4, 107, 108 n. 3, 112, 118 n. 1, 119, 125.
Newark (N. J.), 119.
Newburyport, 105, 210, 212.
Newhall family of Lynn, entrepreneurs of D, 50, 99, 193–196.
Newspaper articles or advertisements quoted or referred to, 29, 168, 169, 188–192, 245 n. 1.
Nichols, John Brooks, adapter of Howe sewing machine to boot and shoe industry, 94.
Norfolk county, a centre for boot and shoe industry, 67, 81, 99, 208, 211, 212.
North Adams, a shoe town in F, 143, 150.
North Bridgewater, see Brockton (present name of town).
North Brookfield, see Brookfield.

PARTS to a shoe, 3.
Peabody, 190 n. 1.
Pennsylvania in F, 118, 154.

282 INDEX

Philadelphia,
  in Ha, 10, 190.
  in D, 28, 29 n. 3, 32, 33, 72 n. 2, 86, 142 n. 1.
  in F, 112, 118, 119, 144.
Plymouth county, a centre for boot and shoe industry, 67, 99, 113, 209, 211, 212.
Political history of four Massachusetts shoe centres, early, 176–183.
Portland (Me.), 33.
Portsmouth (N. H.), 30, 188, 193, 195, 197, 199.
Pratt, J. Winsor, leather shoe-string manufacturer of Randolph, 90, 91.
Prouty, Isaac, the rich cobbler-farmer of Spencer, 136, 137.
Putting-out System, see Stages of Production, Domestic stage.

QUABAUG, see Brookfield (present name of town).

RANDOLPH (formerly W. Randolph, and part of Braintree), a shoe town,
  early conditions in, 11, 17–19, 176–180.
  in D, 19, 45, 50, 52, 72, 73, 81–92, 135, 136, 180, 208, 217 n. 1.
  in F, 98–101, 103–109, 110 n. 1, n. 2, 114, 115, 117, 153, 262.
See also Braintree; S. Randolph.
Raynham, a shoe town,
  in D, 87, 92 n. 1.
  in F, 102, 110–112, 114, 115, 139.
Reading, 72, 208, 211.
Reed, George, of Dighton, farmer in Ho, excerpts from papers of, 173–176.
Reed, Harvey and Quincy, of Weymouth and Boston, typical entrepreneurs of late Ha and early D, 13–15, 49, 50.
Religious history of four Massachusetts shoe centres, early, 176–183.
Rhode Island, boot and shoe industry rarely developed beyond Ha in, 35 n. 3.

Roads, lack of, in early days, 177, 178.
Roberts, James A., *New York in the Revolution as Colony and State*, 192, 193.
Robinson and Co. of Lynn, typical entrepreneurs of late D, 79, 80, 219.
Rockland, 61.

SACHS, Hans, German poet-traveller-shoemaker, 130, 133.
St. Crispin, patron saint of shoemakers, 130, 145, 170, 258–262, 264–266.
St. Crispin's Day, 130.
St. Hugh, patron saint of shoemakers, 130, 169, 264, 265.
Salem, as shipping-point in D, 34, 196.
Sampson, Calvin T., of North Adams, entrepreneur of F who first made use of Chinese labor, 143, 150.
San Francisco, 103, 107.
Saugus, see Lynn (present name of town).
Scope of present volume, vii.
Sewing machines, see Stages of Production (D3 and F1), sewing machines.
Sherman, Roger, shoemaker whose trade knowledge helped the Continental budget in the Revolutionary War, 132, 133.
Shakespeare, William, *King Henry V*, 265, 266.
Shoe-blacking, early advertisement of, 30.
Shoe workers, see Human element, shoe workers.
Shooe-maker's Glory, The, an old song, 264, 265.
Shops, see Stages of Production (Ha, D1, D2, D3, F1), shops.
Skinner and Ward, general storekeepers in Brookfield in early D, 55–57.
Songs of shoemaking, 130, 145, 264, 265.
Sources,
  description of, vi.
  difficulties of collection of, v.
  for Ho, 4, 5 n. 2, 173–176.
  for Ha, 173, 184, 185, 188–190.
  for D, 29, 36, 37, 184, 185, 188–192, 194.
  for Human element, 127, 128.

INDEX 283

South Abington, 245.
South Africa, leather imported from, in D, 30.
South America, sale of Massachusetts made shoes in, in D2, 62, 63.
South Bridgewater, 138.
   See also Bridgewater; Brockton (North Bridgewater).
South Randolph, 46.
   See also Braintree; Randolph.
South Weymouth, 62.
   See also Weymouth.
Southworth, Jedediah, of W. Stoughton, farmer-shoemaker of Ha, excerpts from papers of, 183–185.
Spencer, a shoe town, 72, 136, 137, 210.

STAGES OF PRODUCTION in boot and shoe industry, vii.
Bücher's views confirmed, vii.
HOME stage, 3–8, 15–23.
   account books show conditions in, 173–176.
   approximate extent of, 15.
   capital, lack of, in, 25.
   care taken of shoes in, 178 n. 1.
   characterized, 8.
   economic conditions during, 15–23.
   lasts in, 5.
   leather,
      kinds in use in, 5–7.
      preparation of, in, 4, 6.
      tanning of, in, 6, 7, 18, 182.
   market, lack of, in, 8.
   Phase 1 (purely home-made boots), 5, 6.
   Phase 2 (itinerant cobblers' work), 5–7.
      See also Human element, shoe workers, itinerant.
   place of, in evolution of industry, vii.
   processes of shoemaking in, 3, 4.
   rarity of indulgence in outside work by farmers in, 179.
   summary of, 7, 8.
   tanning, see Stages of Production, Ho, leather, tanning of.
   tools required in, 3–5.
   transition to Ha, 8, 9, 18, 20, 22, 32 n. 1.
   typical of frontier life, 15, 16.

HANDICRAFT stage, 3, 4, 8–23.
   account books show conditions in, 173, 184, 185.
   approximate extent of, 15.
   bespoke work in, 8–11, 19, 25, 128.
   capital, late rise of, in, 25.
   characterized, 8, 24.
   competition, spur of European, in, 10.
   custom work in, 134, 138.
      See also Stages of Production, Ha, extra sale work.
   customers required to go to shoemakers' shops by Massachusetts Court in 1644, 134.
   division of labor, see Stages of Production, Ha, specialization.
   early appearance and disappearance of, in thickly settled communities, 9.
   economic conditions during, 15–23.
   extra sale work in, 11–15, 188, 189, 191.
      closely connected with D, 24.
      first appearance of, 12.
      growth of practice of, 12, 13.
      influence of old English customs on, 128.
      late development of, in Brockton, 20; in Middleboro, 134.
      methods of sale of, 12.
      profits of, 13.
      standards of work low for earliest, 13.
      transition to, 11, 25.
   "fair price" in, 9.
   height of, 9.
   leather,
      importing of, in, 7.
      plentiful, in, 171.
      prices of, in, 173, 186.
      provided by customers occasionally in, 172.

STAGES OF PRODUCTION (*cont.*)
  HANDICRAFT stage (*continued*)
    tanning of, in, 7, 18, 20.
    See also Stages of Production, Ha, stock.
    legislative restrictions in, 9.
    markets,
      development of, in, 8.
      directly dealt with by shoemaker in, 24.
      uncertain in, 12.
    materials, *see* Stages of Production, Ha, stock.
    newspaper advertisements for the shoe industry in, 188–190.
    Phase 1 (bespoke work — *which see*), 8–11.
    Phase 2 (extra sale work — *which see*), 11–15.
    place in evolution of industry, vii.
    prices of shoes in, 173, 185, 187.
      kept high by organized shoe workers, 171.
    repairing, charges for, in, 185.
    shops, small, 134, 138.
      customers required to go to, in *1644* by Massachusetts Court, 134.
      organization of, in, 10.
    specialization, lack of, in, 11.
    standards of work in, 8, 10, 11, 13.
    stock, prices of, in, 186, 187.
    See also Stages of Production, Ha, leather.
    tanning, *see* Stages of Production, Ha, leather, tanning of.
    ten-footers, *see* Stages of Production, Ha, shops, small.
    tools in,
      compared with mediæval tools, 10.
      prices of, 186, 187.
    transition from Ho, 8, 9, 18, 20, 22.
    transition to D1, 14, 19, 20, 22, 24, 40, 54, 57, 58.

  DOMESTIC stage, 24–95.
    appearance of, in England, 26, 27.
    characterized, 24.
    market,
      indirectly dealt with by shoemaker in, 24.
      national, as a cause of its appearance, 27.
    Phase 1, 24–41.
      account books show conditions in, 37, 184, 185, 194.
      barter conditions in, 31, 32.
      booksellers as shoe retailers and entrepreneurs in, 33.
      capital,
        attracted by widening and assurance of market in, 41.
        rise of, in, 25–28.
      capitalist, *see* Stages of Production, D1, entrepreneurs.
      characterized, 25.
      children's shoes, Lynn's specialty in, 39, 40.
      containers used for exporting shoes in, 30.
      correspondence of shoe merchants shows conditions in, 36.
      credits long in Southern sales in, 141.
      early opposition to, in England, 27.
      entrepreneurs,
        appear for first time, 24.
        not necessarily shoemakers, 25, 33, 37.
        reasons for non-shoemakers becoming, 37.
      export trade, *see* Stages of Production, D1, West Indies.
      extent of, 24, 25.
      financial conditions in, 31, 32.
      importation of foreign-made shoes in, 28–30, 188.
      introduction of, into Nantucket, 137.
      inventory of typical retail store in, 197.
      leather,
        advertisements for, in, 31–34.
        dealings in, in, 194, 195.
        importation of, in, 30.

INDEX 285

STAGES OF PRODUCTION (*cont.*)
  DOMESTIC stage (*continued*)
    Phase 1 — leather (*continued*)
      largely domestic, in, 31.
      prices of, in, 197.
      tanning of, in, 190–192.
      See also Stages of Production, D1, stock.
    letters from domestic workers in, 198.
    market,
      capital attracted as it is widened and assured, 41.
      widening of, in, 28.
    materials, *see* Stages of Production, D1, stock.
    men's shoes manufactured in southern and western Massachusetts in, 40.
    middlemen-retailers in, 197–200.
    newspaper advertisements for the shoe industry in, 29, 188–190.
    opposition to, in America, lack of early, 27.
    prices of shoes in, 33, 38, 185, 194, 197–200.
    profits in, 199.
    public vendues in, 30, 189.
    repairing, charges for, in, 185, 199.
    retail trade in, 33, 193–197.
    Revolutionary War,
      exemption of shoemakers from military service during, 192, 193.
      as stimulus to the industry, 29, 39–41, 132.
    shops, small,
      overhead charges slight in, 141.
      sale of, in, 191.
    Southern sales in, 20 n. 2, 34–36, 141, 196.
    standards, influence of foreign, 28, 29.
    stock,
      imported for use in, 28–30.
      prices of, in, 39.
      supplied by entrepreneur in, 24.

See also Stages of Production, D1, leather.
    summary, 40, 41.
    tanneries, newspaper advertisements regarding, in, 190–192.
    tanning, *see* Stages of Production, D1, leather, tanning of.
    tardy development of, in central, northern and western Massachusetts, 31, 34.
    tariff,
      feared by British consul in, 37.
      legislation of *1789* a stimulus, 39–41.
    ten-footers, *see* Stages of Production, D1, shops, small.
    tools,
      imported for use in, 28, 30, 189.
      prices of, as shown by account books, 39.
      supplied by entrepreneur in, 24.
    transition from Ha, 14, 19, 20, 22, 24, 27, 40, 54, 57, 58.
    transition to D2, 44, 45, 58.
    West Indies, sales in, 14, 29 n. 3, 34–36, 144.
    wholesale trade in, 33.
    women's shoes, Lynn's specialty in, 39, 40.

  Phase 2, 42–64.
    account books show conditions in, 44–48, 55, 201–207.
    banking interests of shoe entrepreneurs in, 50.
    barter conditions in, 51, 56.
    in export trade, 62.
    Boston and Worcester Railroad, effect of construction of, 53.
    capital,
      investments of, increased in, 42.
      slow growth of, in, 201.
    capitalists, *see* Stages of Production, D2, entrepreneurs.
    central shop, *see* Stages of Production, D2, shops, central.
    characterized, 42.

STAGES OF PRODUCTION (*cont.*)
  DOMESTIC stage (*continued*)
    Phase 2 (*continued*)
      check to industry due to Panic of *1837*, 64.
      competing industries in, 210–212.
      competition in industry, growth of, in, 42.
      competition of buyers leads to lowering of standards in, 74.
      containers for shipping shoes in, 51, 57, 61, 62.
      credits long in Western and Southern sales in, 141.
      custom work in, 47, 49, 57.
      cutting in central shops in, 44, 49.
      division of labor, *see* Stages of Production, D2, specialization.
      entrepreneurs,
        assist retail firms to get established in, 59.
        Batchellers typical in, 55–59.
        central shop developed by, in, 43, 48, 49, 110 n. 3.
        earliest in Randolph in, 45, 46.
        as general store-keepers in, 45, 46, 48, 49, 55–57, 202 n. 2.
        Littlefield Brothers typical in, 51–53.
        in Natick in, 53, 54.
        not necessarily shoemakers in, 49, 74, 110 n. 3.
        other interests and ventures of, in, 49–51, 57, 63.
      export trade, *see* Stages of Production, D2, South America; West Indies.
      extra sale work in, 47, 49, 56, 57.
      foreign trade methods in, 62.
      general economic conditions during, 54–57.
      inspection,
        difficult before appearance of central shop in, 43.
        not properly made by entrepreneurs not shoemakers, 74.

jobbing in, 58, 59, 61.
land interests of shoe entrepreneurs in, 50.
leather,
  "cabbage" stock in, 43.
  prices of, in, 202.
  tanning of, in, 200, 211, 212.
  usually furnished by entrepreneur in, 43, 47.
  waste of, in, prior to appearance of central shop, 43.
output,
  of central shops in, 46.
  of Massachusetts towns in *1837*, 207–210, 212.
  See also Stages of Production, D2, trade, volume of.
Panic of *1837*,
  hard on children, 136.
  marks close of phase, 42, 50, 54, 58, 63, 64.
phraseology, changes in, in, 44.
poor work, causes of, in, 74.
prosperity just prior to *1837*, 63, 73.
quantity more important than quality in, 42.
rapid work necessary in, 42.
repairing in, 47, 49.
retail firms assisted by entrepreneurs in, 59.
shops,
  central,
    developments of, in, 43, 44, 48, 49, 52, 57, 110 n. 3, 139, 201.
    function of, in, 44.
    needed in connection with inspection in, 43.
    overhead charges slight in, 141.
  small,
    develop into central shops in, 139, 201.
    described, 43 n. 1.
    overhead charges slight in, 141.
    waste in, in, 43, 44.

INDEX

STAGES OF PRODUCTION (*cont.*)
  DOMESTIC stage (*continued*)
    Phase 2 (*continued*)
      South American sales in, 62, 63.
      Southern sales in, 42, 49 n. 2, 50, 51, 59–61, 141, 200, 201, 203, 204.
      specialization in processes in, 42–44.
      standards lowered in, 42.
      supplies sold to shoemakers by entrepreneurs in, 49.
      tanning, *see* Stages of Production, D2, leather, tanning of.
      tariff protection in, 63.
      ten-footers, *see* Stages of Production, D2, shops, small.
      trade,
        extent of, in, 202–204.
        volume of, in, 44, 50, 54 n. 2, 56, 58, 204–207.
        See also Stages of Production, D2, output.
      transition from D1, 44, 45, 58.
      transition to D3, 64.
      transportation facilities in, 50 n. 1, 53, 61.
      West Indies sales in, 42, 50, 51, 59, 61, 62.
      Western sales in, 57–59, 141.
      women's shoes in, 211.

    Phase 3, 65–96.
      account books show conditions in, 79, 80, 213–218, 221–230.
      allied industries in, 89–91, 220.
        See also Stages of Production, D3, containers; counters; cut leather; shoe strings; soles.
      Australian sales in, 81, 83.
      banking facilities in,
        importance of, 67.
        increased demand for, 68.
        payment of cash wages made possible by, 85.
        use of, shown by account books, 219.

      boots, high-legged, specialized in, by certain localities in, 72, 73, 81–84, 114, 115.
      boss contractors in ten-footers in, 86.
      Boston and Worcester Railroad in, 216.
      brogans specialized in by certain localities in, 80, 81, 114, 139.
      "Californians," *see* Stages of Production, D3, boots.
      capital in,
        attraction of, 68.
        regaining of, following panic of *1837*, 66.
        small amount only needed to equip a central shop, 79.
      capitalists, *see* Stages of Production, D3, entrepreneurs.
      central shops, *see* Stages of Production, D3, shops, central.
      competition between manufacturers in,
        economies of stock and labor the result of, 90.
        expert supervision of product the result of, 71, 72, 94.
        improvement in stock and processes the result of, 74, 75, 78.
        specialization the result of, 81, 94, 95.
      containers for shipping shoes in, 70, 80, 216.
        manufacture of, as a separate industry, 77, 78.
      counters, cutting of, as a separate industry, 90.
      credits long in Western and Southern sales in, 141.
      "crooked shoes," first appearance of, in, 76.
      custom as determining location of boot and shoe industry in, 67.
      customers, different classes of, as cause of specialization in, 82, 83, 95.
      cut leather industry, 77.

STAGES OF PRODUCTION (*cont.*)
DOMESTIC stage (*continued*)
Phase 3 (*continued*)
distribution of work to domestic workers in, 100.
division of labor, *see* Stages of Production, D3, specialization.
economies of boot and shoe manufacture, 89–92.
entrepreneurs in,
advantageous position of those surviving panic of *1837*, 65.
boot and shoe industry now their main interest, 68.
new type of, 66, 68.
partnership between shoemaker and capitalist common, 71, 139.
factory, genesis of, 86.
"freighters," agents who distributed to domestic workers, 70.
"gang," a factor in shoemaking, 86.
grading, increased stress on, in, 75, 83.
heels, use of, in, 93.
imitation as determining location of boot and shoe industry, 67.
jobbers in, 83.
lasts in, 91, 93, 217, 218.
leather in,
prices of, 115 n. 3, 214, 226, 227.
sale of scrap, for shoe string manufacture, 90, 91.
localities, specialization of industry in certain, 66–68.
machines in,
development of, 78, 80, 92, 95, 96, 139.
few considered necessary, 79.
relation of, to economizing of time and labor, 92, 95, 96.
review of use of, 93–96.
valuation of, in typical shop, 219.
See also Stages of Production, D3, sewing machine; sole-cutter; stripper.
manufactory, 94 n. 3.
See also Stages of Production, D3, shops, central.
markets, new, growth of, in, 66, 81, 82, 94, 95.
materials, *see* Stages of Production, D3, stock.
models, new, appearance of, in, 75, 82, 83, 94, 95.
output, 69, 82.
See also Stages of Production, D3, trade, volume of.
panic of *1837*, revival of industry after, 65, 66.
patterns for cutting leather in, 75, 76, 93.
payment by jobbers, method of, in, 69.
piece-work in homes and ten-footers in, 85, 86.
prices of shoes in, 81, 84, 213–216, 221–225.
ready-made shoes, growth of vogue of, in, 81, 82, 95.
sewing machine in, 78, 93–96.
shoe strings, leather, cutting of, in,
as a separate industry, 90, 91.
as a side line, 224, 225.
shops in,
central,
how used, 71, 72, 94.
increasing importance of, 80, 81.
overhead charges slight in, 141.
processes taking place in, 85, 88.
small amount of capital needed to equip, 79.
small, 71, 100, 218.
overhead charges slight in, 141.
processes taking place in, 85, 86, 88, 89.
rented to shoemakers, 86.
valuation of typical, 214.

INDEX

STAGES OF PRODUCTION (*cont.*)
DOMESTIC stage (*continued*)
Phase 3 (*continued*)
    sizes begin to be carefully noted, 228.
    sole-cutter, 76, 77.
    soles, cutting of, as a separate industry, 90.
    Southern sales in, 70, 72 n. 3, 75, 81, 82 n. 2, 95, 110 n. 3, 139, 141.
    specialization in,
      geographical, 66–68.
      in processes,
        competition as cause of, 81, 94, 95.
        differences in classes of customers as cause of, 82, 83, 95.
        growth of, 72, 73, 80, 81, 94.
        shown by list of occupations, 91.
      in ten-footers, 86.
    standardization in,
      need of, 73, 74.
      growth of, 72–78, 92, 95, 144.
    statistics of manufacture in, 73, 82.
    stock in,
      economy in use of, necessary, 90.
      greater cost of higher grades, 84.
      importance of, 75.
      inventory of, in typical shop, 219.
      purchases of, 225–227.
      See also Stages of Production, D3, leather.
    stripper, a leather-cutting machine in, 76, 77.
    styles, more follow the panic of *1837*, 144.
    summary of, 94–96.
    supervision of labor, rise and growing importance of, in, 71, 72, 92, 94.
    tanneries, importance of nearness to, in, 67.
    tariff in, 79.
    technical progress in, 83.
    ten-footer, *see* Stages of Production, D3, shops, small.
    time allowed to domestic workers unlimited, 100.
    time-saving, tin sole-patterns and, 76.
    tools in,
      century-old models still exclusively used, 93.
      importance of, 75.
      prices of, 227.
    toolmaking in connection with boot and shoe industry in, 67.
    trade, volume of, 69, 84, 92.
      See also Stages of Production, D3, output.
    transition from D2, 64.
    transition to F1, 96, 114–116.
    transportation facilities, importance of, in, 67, 70, 71 n. 1, 83, 84.
    West, expansion of, effect on boot and shoe industry in, 82, 83, 95.
    West Indies sales in, 75, 110 n. 3, 144.
    Western sales in, 72 n. 3, 81, 83, 95, 141.
    women's shoes specialized in by certain localities, 72, 73, 78, 81.
    work, irregular amounts of, in, 229, 230.
    working days per year in, 87.
    workmanship, more precision in, following panic of *1837*, 144.

  place in evolution of industry, vii.
  tardy development of, vi n. 1.

FACTORY stage,
  early opposition to, in England, 27.

Phase 1, 97–126.
  account books show conditions in, 116, 230–244, 246–255.
  allied industries in, 123.

290 INDEX

STAGES OF PRODUCTION (*cont.*)
  FACTORY stage (*continued*)
    Phase 1 (*continued*)
      army shoes for Civil War in, 117, 118, 124, 245, 246.
      Australian sales in, 98–101, 103, 107, 110 n. 1, 267.
      banking facilities in, 102, 106, 230, 231.
      Boston and Albany Railroad in, 123.
      businesses shut down or move because of KSC activities in, 150, 153, 154.
      capital in,
        circulating, less important in this period, 121, 141.
        fixed, dangerously large amounts necessary, 103, 110, 121, 141.
        more required, 124.
        total investment of, in *1860*, 112, 113.
        use of, for factory buildings, 110.
      capitalists, *see* Stages of Production, F1, entrepreneurs.
      central shop, *see* Stages of Production, F1, shops, central.
      characterized, 97, 98, 123, 124.
      Chicago strike of KSC in, 151, 152.
      Civil War during,
        effect of, on shoe industry, 98, 109–118, 124.
        use of McKay machines in, 245, 246.
      competition between manufacturers,
        increasing importance of, 122.
        leads to emphasis on good workmanship and speed, 109.
      Congress shoe in, 101, 111.
      containers for shipping shoes in, 252, 255.
      contract system of wages, trial of, in Brookfield in, 122.
      coöperative manufacture advocated by KSC in, 146, 147.
      delay in appearance of, in Brookfield, 114–116.
      depression due to panic of *1857*, 103–109.
      distribution of business in, 238–244.
      economies, necessity for, a cause of F, 98.
      entrepreneurs in,
        influenced by outside investments, 262.
        more and more from outside trade, 141.
      European trade in, 104 n. 4.
      factories in, 119–121, 140.
        appearance of, not cause of F, 98, 124.
        buildings, development of, 110–112, 139; involve large investments, 141.
        increase of work in, 100, 101.
        output in, size of, 110 n. 1.
        rise of, 97–101, 123, 124.
        size of, indicated by accounts, 246–255.
      failures in, 105, 108, 109, 125.
      fixed improvements in, 103.
      full development of stage in Lynn by *1865*, 120.
      "good low-priced shoe" developed in Brockton in, 118, 119.
      large scale production, adaptation to needs of, 100, 101.
      lasts in, 233, 234, 255.
      leather in,
        hemlock tanned, 118.
        oak tanned, 118.
        prices of, 104–107, 118, 234.
        "Union," 118.
      machines in, 140, 251.
        "abuse" of, attacked by KSC, 146.
        importance of, in connection with investment, 141.
        installation of, not cause of F, 98.
        make use of unskilled labor especially feasible, 143.

INDEX

STAGES OF PRODUCTION (*cont.*)
  FACTORY stage (*continued*)
    Phase 1 — machines in (*continued*)
      use of, made necessary by competition, 109, 110.
      See also Stages of Production, F1, McKay machine; sewing machine; Wheeler and Wilcox machine.
    materials, *see* Stages of Production, F1, stock.
    McKay machine in, 96, 98, 101, 110, 111, 117–119, 121, 124, 143, 156, 254, 255.
      history of invention of, 245, 246.
      royalty system on, 121, 122.
    McKay sewed shoe not superior to custom shoe in, 143 n. 1.
    manufactory, *see* Stages of Production, F1, factories.
    orders, increase in size of, in, 99.
    output in, 110 n. 1, 119 n. 2.
      in *1860*, 112, 113.
      uniformity of, a desideratum leading to F, 98.
      See also Stages of Production, F1, trade, volume of.
    outside investments, influence of, on entrepreneurs in, 262.
    "overproduction" in, 103, 105, 106.
    panic of *1857*,
      general effect on boot and shoe trade of, 103–109.
      recuperation following, 109–111.
    panic of *1873*, 125.
      effect of, on KSC, 156.
    power in,
      introduction of, 124.
      use of, not cause of F, 98, 119, 120.
    prices of shoes in, 232, 233, 235.
    problems that came with, 141, 142.
    processes in manufacture in Lynn in *1880*, 120, 121.
    profits, growth of, in, 99.
    prosperity in, 103.
    putting-out in, 236–239.
    recuperation following panic of *1857*, 109–111.
    sewing-machines in, 100, 101, 109, 120 n. 1.
    shops in,
      central, number of, in *1860*, 112, 113.
      small, decrease in use of, 109, 120.
      use of term for "factory," 101, 106 n. 4.
      See also Stages of Production, F1, factories.
    Southern sales in, 40, 98, 104 n. 4, 106, 107, 109, 112, 114, 238, 239.
    specialization in,
      geographical, 124.
      in processes, increase of, 124;
        in Lynn in *1880*, 120, 121;
        renewed following panic of *1873*, 156.
    speculation in, 103, 107, 125.
    speed in production in,
      adaptation to needs of, 100, 101.
      emphasis on, 109, 110.
    stock in,
      cutting up of, 103, 105.
      development of, 123.
      large supplies of, used, 124, 141.
      purchases of, 254, 255.
      total value used in *1860*, 112, 113.
      See also Stages of Production, F1, leather.
    styles of shoes, great variety of, 106, 125, 144.
    summary of, 123–126.
    supervision at the central shop, chief characteristic of, 97, 98, 123, 124.
      needed to secure better work and to save waste, 109.
    ten-footer, *see* Stages of Production, F1, shops, small.

STAGES OF PRODUCTION (*cont.*)
  FACTORY stage (*continued*)
    Phase 1 (*continued*)
      transition from D3, 96, 114–116.
      transition to F2, 124, 125, 156.
      transportation in,
        account books show costs of, 123, 249, 250, 253–255.
        improvement of facilities of, 101–103, 116.
      methods of, 238–246.
      trade, volume of, in, 104, 107, 108, 122, 123, 238–244, 246.
      See also Stages of Production, F1, output.
      West Indies sales in, 111.
      Western sales in, 98–101, 103, 106, 107, 109, 112, 114, 240–244.
      Wheeler and Wilcox machines in, 140.
      workmanship in,
        emphasis on quality of, 109.
        more precision in, following panic of *1857*, 144.

    Phase 2,
      capital, development of, in, 156.
      characterized, 125, 156.
      factory, modern, organization of, described, 159–167.
      Goodyear welting machine in, 96, 124, 125, 160, 165, 167.
      Goodyear welt shoes characteristic of, 156.
      leather, use of, in modern shoemaking in, 159–163, 165, 166.
      machines, great use of, in modern factories in, 159–167.
      See also Stages of Production, F2, Goodyear welt machine; McKay machine.
      McKay machine, use of, in, 160, 161, 164, 167.
      processes on shoes in a modern factory in, 159–167.
      repairing, modern machines for, in, 168, 169.

      specialization in, 3, 159–167.
      summary of, 125, 126.
      transition from F1, 124, 125.
      welting in, 159, 164, 167.
      See also Stages of Production, F2, Goodyear welt machine.

    place in evolution of industry, vii.

Stoneham, 72, 208.
Storekeeping a frequent activity of shoe entrepreneurs, 45, 46, 48, 49, 55–57, 202 n. 2.
Stoughton, a shoe town,
  in Ha, 132, 180.
  in D, 72, 90, 208.
  in F, 99.
  See also Avon (E. Stoughton); W. Stoughton.
Strong's factory at Randolph, 99 n. 1, 100 n. 2, 101.

TEMPLE, Josiah H., *History of North Brookfield*, 182 n. 1, 200 n. 1, 201 n. 1, n. 2, n. 4.
Tobin, John F., General President of the Boot and Shoe Workers' Union, quoted, 146 n. 1.
Tools, shoemaking, mediæval, 169. *See also* Stages of Production (Ho, Ha, D1, D3), tools.
Towns, development of new from old, 17.
Trades in New England in *1650*, contemporary account of, 170, 171.
Twitchell, Henry Emmons, of Brookfield, entrepreneur of F, 114, 115 n. 2, 116, 236–244.

UNITED Shoe Machinery Co. gives repair outfit to U. S. Army, *1917*, 168.
United States Bureau of Labor Statistics, *Bulletin*, 159–167.
Unwin, George, *Industrial Organization in the 16th and 17th Centuries*, 26, 27.

INDEX

WAGES, *see* Human element, wages.

Wales, Jonathan, of Randolph, shoe merchant in San Francisco, in D, 89, 98, 99, 101.

Ward, Oliver, of Grafton, inaugurator, as early entrepreneur, of D in Brookfield, 22, 55, 57, 200–202.

Wendell family of Portsmouth (N. H.), entrepreneurs of early D who were not shoemakers, 37, 193–200.

West Indies,
leather imported from, in D, 30.
sale of Massachusetts made shoes in,
in D1, 14, 29 n. 3, 34–36, 144.
in D2, 42, 50, 51, 59, 61, 62.
in D3, 75, 110 n. 3, 144.
in F1, 111.

West Randolph, *see* Randolph (present name of town).

West Stoughton, 183.
See also Avon (E. Stoughton); Stoughton.

Weymouth, a shoe town,
early conditions in, 9, 13, 14, 137, 178.
in D, 63, 72, 100 n. 1, 208, 211.
in F, 101.
See also South Weymouth.

White, Samuel, of Randolph, expert pegger of late D, 135, 136.

White, Timothy, of Nantucket, teacher-shoemaker, of Ha, 12 n. 2, 173.

Wilson, Henry, of Natick, entrepreneur of Natick, later vice-president of the United States, v n. 2, 68–71, 132, 213–218.

Winks, William, *Lives of Illustrious Shoemakers*, 128, 256 n. 1, 258 n. 1, 261 n. 1, 262 n. 1.

Woburn, 72, 190, 208, 211.

Wolcott, John B., of Natick, entrepreneur of D, 53, 69 n. 3, 220.

Women as shoeworkers, *see* Human element, women.

Wooldredge, John, of Lynn, entrepreneur of late D, 78, 120.

Woolen industry, development of, in England compared with development of American boot and shoe industry, 26, 27.

Worcester, a former shoe town,
in D, 21, 33, 34, 72, 200, 210.
in F, 147, 150.

Worcester county, a centre for boot and shoe industry, 21, 67, 81, 113, 209, 210, 212.

www.ingramcontent.com/pod-product-compliance
Lightning Source LLC
Chambersburg PA
CBHW021144160426
43194CB00007B/680